Levulinic Acid

Levulinic Acid

A Sustainable Platform Chemical for Value-Added Products

Claudio J.A. Mota
Ana Lúcia de Lima
Daniella R. Fernandes
Bianca P. Pinto
Federal University of Rio de Janeiro
Institute of Chemistry
Rio de Janeiro
Brazil

WILEY

Registered Offices
John Wiley & Sons, Inc., 111 River Street, Hoboken, NJ 07030, USA
John Wiley & Sons Ltd, The Atrium, Southern Gate, Chichester, West Sussex, PO19 8SQ, UK

For details of our global editorial offices, customer services, and more information about Wiley products visit us at www.wiley.com.

Wiley also publishes its books in a variety of electronic formats and by print-on-demand. Some content that appears in standard print versions of this book may not be available in other formats.

Library of Congress Cataloging-in-Publication Data:

Names: Mota, Claudio J. A., author. | Lima, Ana Lúcia de, author. | Fernandes, Daniella R., author. | Pinto, Bianca Peres, author.
Title: Levulinic acid : a sustainable platform chemical for value-added products / Claudio J.A. Mota, Ana Lúcia de Lima, Daniella R. Fernandes, Bianca P. Pinto.
Description: Hoboken, NJ : Wiley, 2023. | Includes bibliographical references and index.
Identifiers: LCCN 2022038112 (print) | LCCN 2022038113 (ebook) | ISBN 9781119814665 (cloth) | ISBN 9781119814689 (adobe pdf) | ISBN 9781119814696 (epub)
Subjects: LCSH: Ketonic acids.
Classification: LCC QD341.K2 M78 2023 (print) | LCC QD341.K2 (ebook) | DDC 547/.036–dc23/eng20221123
LC record available at https://lccn.loc.gov/2022038112
LC ebook record available at https://lccn.loc.gov/2022038113

Cover Design: Wiley
Cover Images: © tchara/Adobe Stock Photos; © Peggy_Marco/pixabay; © ybernardi/pixabay; © stokpic/pixabay; Molecular Structures: Courtesy of Daniella R. Fernandes

Set in 9.5/12.5pt STIXTwoText by Straive, Chennai, India
Printed and bound by CPI Group (UK) Ltd, Croydon, CR0 4YY

C9781119814665_161222

Contents

About the Authors

Claudio J.A. Mota holds a degree in chemical engineering from the Federal University of Rio de Janeiro (UFRJ), where he also obtained his PhD in chemistry. He is full professor of chemistry and chemical engineering, as well as director of the Institute of Chemistry of the UFRJ. He is a research fellow from CNPq and state scientist from FAPERJ. He is member of the Brazilian Chemical Society (SBQ), Brazilian Catalysis Society (SBCat), American Chemical Society (ACS), and fellow of the Royal Society of Chemistry (RSC). He was awarded with the TWAS Prize, given by the Mexican Academy of Science, the Technology Prize from the Brazilian Association of the Chemical Industries (ABIQUIM), the Innovation Prize of SBQ, and the Simão Mathias Medal, the highest honor of the SBQ. He is author of more than 160 scientific publications, among articles, patents, and books. He participates in the editorial boards of the *Journal of CO_2 Utilization, Journal of Catalysis and ACS Omega*, having also established several international collaborations. His current research interests are focused on biomass transformation and processes of CO_2 capture and conversion, targeting applications in the fuel and chemical sectors.

Ana Lúcia de Lima received her BA degree (2011) in Chemical Sciences with Technological Assignments in Chemical Sciences from the Federal University of Uberlândia (UFU), Brazil. She received her MSc (2013) and doctoral (2017) degrees from Federal University of Rio de Janeiro (UFRJ), Brazil. She has started postdoctorate study (2017) at LMCP/IMA/UFRJ on development and evaluation of polymeric hydrogels for compliance control in oil reservoirs. At the moment, she is a professor at the Department of Analytical Chemistry at the Institute of Chemistry at UFRJ and a researcher at LARHCO/UFRJ where she researches in the area of mesoporous materials with various applications, such as heterogeneous catalysts with a focus on biomass transformation and adsorbents for CO_2 capture.

Daniella R. Fernandes is an adjunct professor at the Department of Organic Chemistry at the Institute of Chemistry of the Federal University of Rio de Janeiro (UFRJ), Brazil. She holds a degree in chemistry (2001), master's (2004), and doctorate (2009) in chemistry from the same university, all with specialization in petroleum chemistry. She is a permanent professor in the Chemistry Professional Master's Program in the National Network (PROFQUI) at the Institute of Chemistry, UFRJ. She has been working at the LARHCO in the environment, energy, and catalysis areas, especially correlated to biomass valorization, biofuels production and chemicals, and material development for CO_2 capture and utilization.

Bianca P. Pinto graduated in chemistry (2006) from the State University of Rio de Janeiro (UERJ) and obtained her master's (2009) and doctoral degrees (2013) in chemistry from the Federal University of Rio de Janeiro (UFRJ, Brazil). She continued to work at the same university for a postdoctoral stay (2014–2021). Her research focused on the catalytic transformation of levulinic acid into valerolactone and other products, and CO_2 capture and conversion. She is also a cofounder of CarbonAir Energy, carbon capture, and utilization startup. She is currently a substitute professor at the Department of Analytical Chemistry at the UFRJ. She is the author or coauthor of 14 scientific articles in indexed journals, 3 book chapters, and 2 published books.

Preface

The climate changes caused by the use of fossil resources are driving the search for more sustainable energy sources, and the use of renewable raw material for the chemical industry. In this context, biofuels and bioderived products have emerged in the world's scenario as alternatives to decrease the carbon emissions. Bioethanol and biodiesel are commercially produced in many countries, whereas bio-based commodity chemicals, such as ethylene obtained from ethanol, are industrially produced in large scale. The bio-based economy will continue to grow in the twenty-first century, especially that dealing with the valorization of lignocellulosic materials and agriculture residues of less economic value. Therefore, the development of processes for converting lignocellulosic biomass into valuable products and fuels is of great importance.

Levulinic acid (LA) emerges as an important bioderived feedstock, as it may be obtained from sugars employing thermochemical routes. Therefore, its production can be accomplished using great diversity of biomass raw materials, which makes levulinic acid a versatile bio-based platform chemical. Today, it is still considered a specialty chemical and its industrial production is limited, as well as the applications of LA and major derivatives in different sectors. Nevertheless, as the demand for new bio-based products increases, it is expected that levulinic acid may grow in importance, being one of the main bioderived feedstocks for the production of chemicals and biofuels.

This book intends to gather the current knowledge on levulinic acid production and conversion into major derivatives. From the best of our knowledge, this is the first book on the subject, and we hope it can motivate new scientific and technological developments, as well as new uses for levulinic acid and its derivatives. LA is a keto-carboxylic acid; thus, it presents two functionalities that could be exploited in different reactions. Such versatility is not usually found in simple bioderived molecules, making levulinic acid an attractive bio-based platform as pointed out by the US Department of Energy.

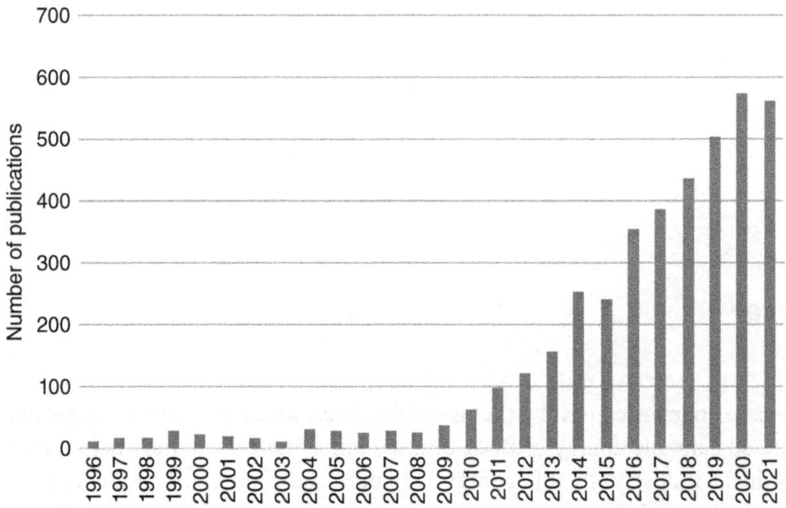

Figure 1 Number of published scientific articles with levulinic acid as keyword.

Figure 1 shows the number of scientific publications with levulinic acid as keyword from 1996 to 2021. The rising interest on the subject is evident, especially after 2010. The need for decreasing the CO_2 emissions in the forthcoming years, to meet the goals of recent UN climate change agreements, will push forward the research on bioderived platforms, such as levulinic acid. Therefore, this book may help scientists and students around the globe in the search of new applications and processes for production and transformation of levulinic acid and its derivatives. We hope that the content would be useful, and the information needed would be easily found in one of the chapters.

Chapter 1 covers the historical context, highlighting the first studies on sugar hydrolysis that may have produced levulinic acid. It also describes the properties and first industrial processes and producers of LA. The chapter highlights the current market and companies that commercialize LA, also discussing the main applications of LA and some derivatives, as well as toxicological issues.

In Chapter 2, the processes of LA production are thoroughly discussed. Different routes, catalysts, feedstocks, and reaction conditions are presented, especially focusing on the production from lignocellulosic biomass materials. The Biofine process is also discussed, as well as the challenges on product separation and purification.

Chapter 3 is dedicated to the organic and inorganic levulinates. The production routes for the levulinate esters are detailed and discussed, together with the current market and main applications. Levulinate esters are versatile chemicals with applications going from the fuel sector to the food industry. The chapter also discusses the production route and main uses of inorganic levulinates, especially sodium and calcium levulinates. Both have uses in the food and pharmaceutical sectors.

The hydrogenation of LA is covered in Chapter 4. The different products that can be obtained from LA hydrogenation are discussed, together with information on catalysts, reaction pathways, and conditions. A particular emphasis is given to γ-valerolactone (GVL), which has many potential applications. However, other hydrogenated derivatives, such as 1,4-pentadediol (1,4-PD) and 2-methyl tetrahydrofuran (2-MTHF), are also discussed together with their main uses.

Chapter 5 is dedicated to reactions on the carbonyl group. Formation of ketals, especially the glycerol levulinic acid/ester ketal (GLEK), is discussed, together with the main potential applications. The chapter also discusses the process of LA conversion to succinic acid and the production of δ-aminolevulinic acid (DALA), as well as their main uses. The reaction of levulinic acid/esters with phenol derivatives is also highlighted as a bioderived strategy to replace bisphenol-A (BPA).

Finally, Chapter 6 shows examples of LA production and uses in a sugar cane biorefinery. Production of ethyl levulinate and GVL is discussed, highlighting the integration of these chemicals with other products of the sugar cane biorefinery. In addition, integration of the biodiesel and LA fabrication chains was discussed, aiming at the production of GLEK.

We believe that the book may be a good and updated source of reference to students, scientists, and professionals in the academia, industry, and government, also motivating new technological discoveries and commercial uses of LA and its derivatives. Our aim in writing this book was to provide concise information on LA and its derivatives, covering technical aspects and major uses, together with relevant references for further consult of the interested reader.

Rio de Janeiro, May 2022

Claudio J.A. Mota
Ana Lúcia de Lima
Daniella R. Fernandes
Bianca P. Pinto

1

Levulinic Acid – History, Properties, Global Market, Direct Uses, Safety

1.1 History and Properties

The first evidence for the formation of levulinic acid was obtained from the treatment of sugars with dilute acid solutions. The Italian-French Pharmacist Faustino Jovita Malaguti (Figure 1.1) reported, in 1835 [1], the treatment of sucrose with boiling diluted acid solutions, being able to identify formic acid and other ammonia-soluble compounds. In 1840, the Dutch chemist Gerardus Johannes Mulder reported the treatment of fructose with hydrochloric acid and was able to isolate acidic compounds [2]. Although these two scientists did not explicitly identify levulinic acid among the products, they were the first to conduct experiments in which levulinic acid would be formed. Nevertheless, the first identification of levulinic acid from the acid treatment of sugars was reported by Tollens in 1875 [3].

Levulinic acid (LA) is a keto-carboxylic acid bearing carbonyl and carboxyl groups in its structure (Figure 1.2). Therefore, the double functionalization makes it an interesting chemical for multiple purposes. Reactions involving levulinic acid are known since the 1870s, but the development of commercial processes and uses were not significant until the 1940s, mostly because of the high costs of the raw materials at the time, high capital costs, and low yields. With the use of cellulose-based feedstocks [4], the production of levulinic acid became more attractive motivating its general use.

Levulinic acid is the 4-oxo-pentanoic acid according to the International Union of Pure and Applied Chemistry (IUPAC) nomenclature. It may also be regarded as the 4-oxo-valeric acid. In the pure form, it is a solid that melts at 30–33 °C and boils at 245–246 °C at atmospheric pressure. Table 1.1 shows some selected properties of levulinic acid.

The first commercial production of levulinic acid dates from the 1940s when the A.E. Staley Manufacturing Company started a bath production using starch as raw material. The process [5] involves the initial mixing of proper amounts of starch

Levulinic Acid: A Sustainable Platform Chemical for Value-Added Products, First Edition.
Claudio J.A. Mota, Ana Lúcia de Lima, Daniella R. Fernandes, and Bianca P. Pinto.

Figure 1.1 Faustino Jovita Malaguti (1802–1878) (https://www.redalyc.org/jatsRepo/1816/181662291006/html/index.html). Source: CITMATEL.

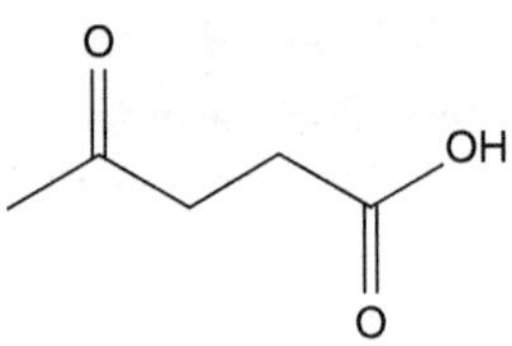

Figure 1.2 Structure of levulinic acid.

and diluted hydrochloric acid at 100 °C in a preheater system. Then, the reaction mixture was autoclaved at 175–215 °C for a determined period. The effluent was neutralized with soda ash and the humins, which are insoluble by-products, were separated by filtration, whereas water and formic acid, formed as a by-product, were evaporated from the solution and the sodium chloride by centrifugation. Levulinic acid was obtained, like a light-colored liquid, upon vacuum steam distillation (Figure 1.3).

Starch is a biopolymer made of hexoses, mainly glucose, that may be obtained from wheat, potato, and oat among numerous other crops. The chemical pathway from C6 sugars to levulinic acid is depicted in Scheme 1.1. The acid medium dehydrates the carbohydrates to intermediate compounds, like hydroxymethyl furfural (HMF), which can then be converted into levulinic acid also releasing a molecule of formic acid in the process.

Table 1.1 Selected properties of levulinic acid.

Property	Value
Chemical formula	$C_5H_8O_3$
Molecular weight	116.11 g
CAS number	123-76-2
Chem Spider ID	11 091
Specific mass	1.1340 g ml^{-1} (25 °C)
Melting point	30–33 °C
Boiling point	245–246 °C
Flash point	137 °C
Refractive index	1.439
pKa	4.65 (water at 25 °C)
Solubility in water	675 g l^{-1} (20 °C)
Solubility in ethanol	Soluble
Solubility in hydrocarbons	Insoluble
Aspect	White to clear yellow solid or liquid
LD_{50} in rats (oral)	1850 mg kg^{-1}

In the 1950s, the Quaker Oats Company developed a process to produce levulinic acid based on furfuryl alcohol as raw material. The company started producing furfural from sugarcane bagasse or corn cobs in 1922 [6, 7], upon heating the biomass in an aqueous solution of sulfuric or phosphoric acid. Furfural is an aromatic aldehyde mostly used in the production of resins at that time. In 1934, the company began the production of furfuryl alcohol from the high-pressure hydrogenation of furfural in Memphis, Tennessee, United States [8]. Then, a continuous process was developed, achieving 99% conversion of furfural in furfuryl alcohol with the use of copper-supported catalysts [9]. The production of levulinic acid from furfuryl alcohol began in 1957 and ended in 1972. The process involved the heating of furfuryl alcohol in the presence of aqueous hydrochloric acid, but small alcohol, such as methanol and ethanol, could also be employed affording the respective levulinate esters [10]. In the beginning, the company did not have enough uses for the levulinic acid produced. In 1959, Quaker Oats released [11] a contest for someone to bring *a big idea on a commercial use of levulinic acid* (Figure 1.4).

The overall chemical pathway to produce levulinic acid from C5 sugars is shown in Scheme 1.2. The pentoses are initially dehydrated to furfural in the acidic medium, usually hydrochloric acid. Then, in a second process, furfural

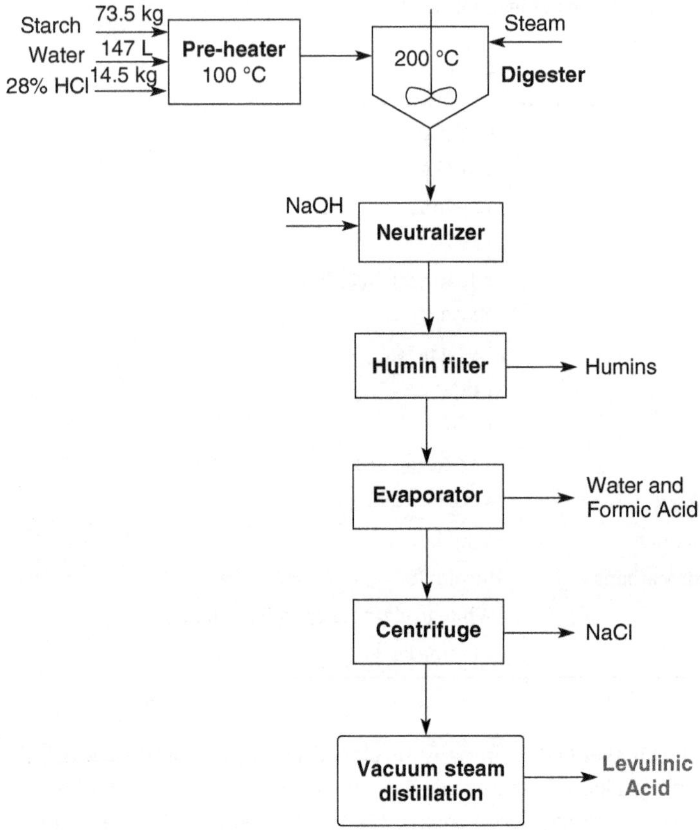

Figure 1.3 Flow diagram of levulinic acid production process.

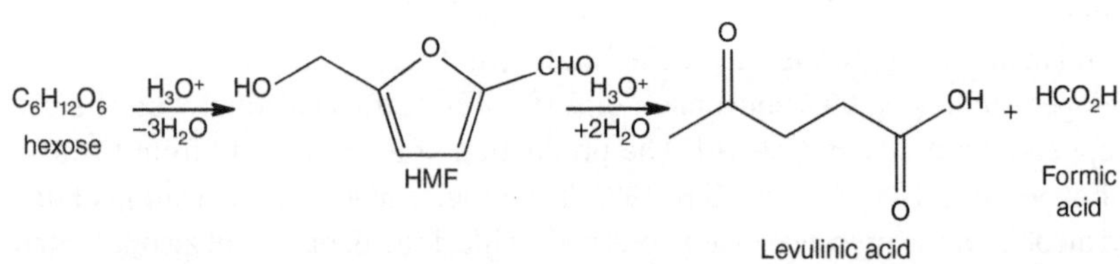

Scheme 1.1 Schematic reaction pathway for the production of levulinic acid from hexoses.

is hydrogenated to furfuryl alcohol, which is further converted to levulinic acid upon acid-catalyzed hydrolysis.

A biotechnological route to levulinic acid involving multiple steps has also been suggested [12]. It involves the fermentation of sugars, like glucose and fructose,

Figure 1.4 Advertisement of the contest for uses of levulinic acid in 1959. Source: Reproduced with the permission of the American Chemical Society.

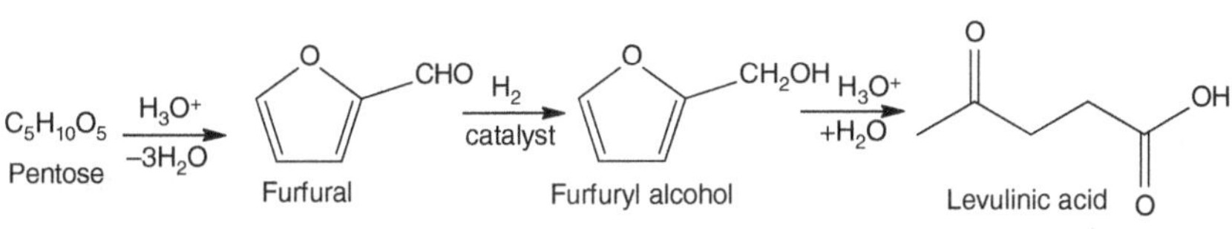

Scheme 1.2 Schematic reaction pathway for the production of levulinic acid from pentoses.

to pyruvic acid as the first step (Scheme 1.3). The next steps may also involve biocatalysts. For instance, acetaldehyde can be produced from pyruvic acid with the use of pyruvate decarboxylase. In the same way, aldolases may be employed in the aldol condensation step. Dehydration of the 2-hydroxy-4-oxo-pentanoic acid, followed by the selective hydrogenation of the intermediate to levulinic acid, may be

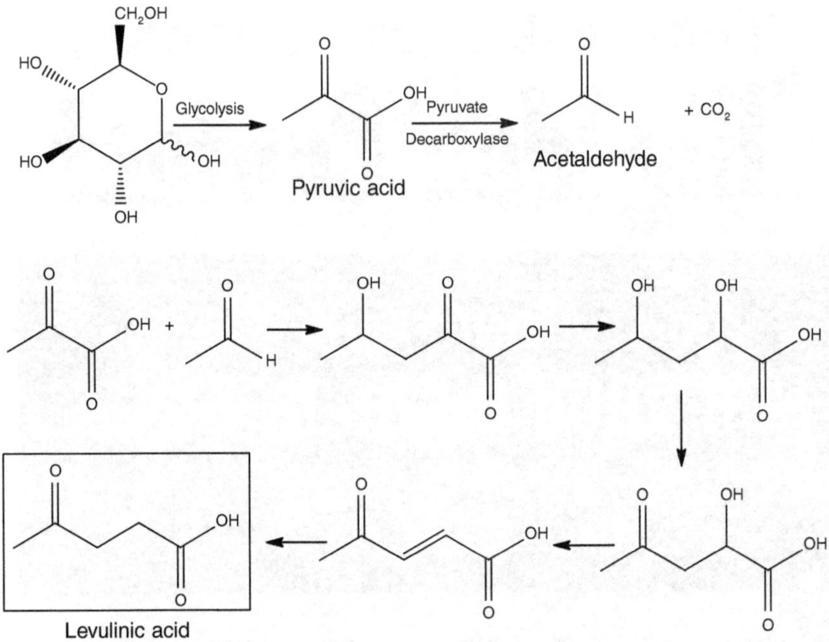

Scheme 1.3 Biotechnological route for the synthesis of levulinic acid.

carried out either through biocatalysis or homogeneous/heterogeneous catalysis. Although technically feasible, this route has not been employed industrially, probably because of the numerous steps, which may require specific reaction conditions and separation procedures, significantly increasing the production costs.

A fossil-based route to levulinic acid has also been developed [13]. The raw material is maleic acid, which is normally obtained from oxidation of benzene [14] or butenes [15]. However, recent developments point out possible routes from renewable feedstocks [16]. Scheme 1.4 shows the steps, which involve the decarboxylation of acetyl succinate.

The DSM company in Linz, Austria, has developed a small-scale process to produce 3 tons per day of levulinic acid from maleic acid, with an overall yield of about 80%. Nevertheless, the company has moved the production of levulinic acid from bio-based raw materials, discontinuing the fossil-based route.

Levulinic acid is a bifunctional molecule, having a keto-carbonyl and a carboxylic acid group. Therefore, it is expected to present the chemical properties of ketones and carboxylic acids. Scheme 1.5 shows the most common transformations of the levulinic acid molecule.

The levulinate esters find applications as fuel additives [17], as well as fragrancies. Inorganic levulinate salts are also important specialty chemicals; sodium

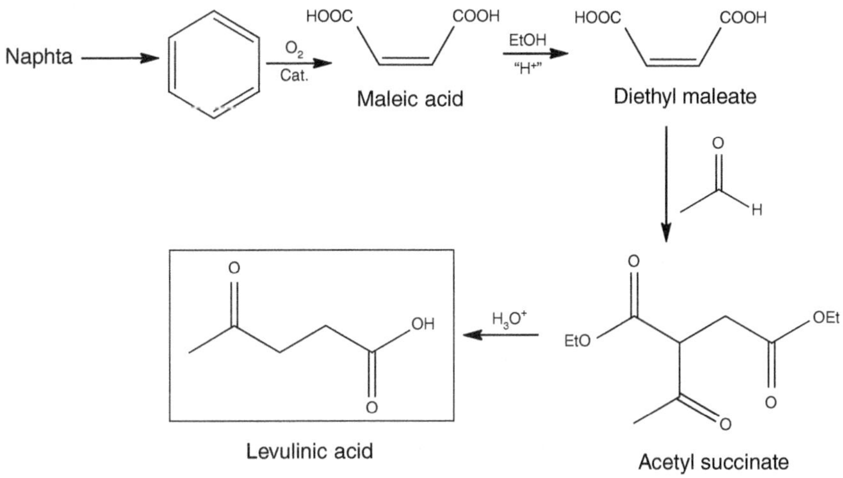

Scheme 1.4 Synthesis of levulinic acid from fossil sources; use of maleic acid as feedstock.

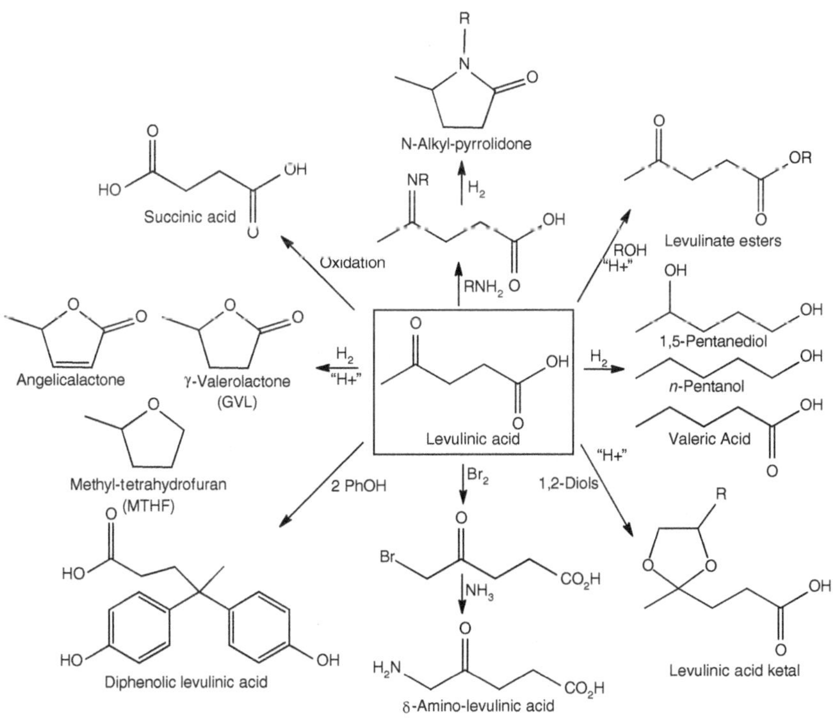

Scheme 1.5 Main levulinic acid transformations.

levulinate is used in the cosmetic and food industry as preservative [18], whereas calcium levulinate is used in pharmaceutics and as a supplementary source of calcium. Hydrogenation may lead to different products. The γ-valerolactone (GVL) is normally produced upon the hydrogenation of levulinic acid over bifunctional catalysts, having a metallic and acidic function. Angelica lactone and methyl-tetrahydrofuran (MTHF) may also be produced depending on the catalyst and reaction conditions. GVL and MTHF are potential fuel additives [19] and green solvents [20], whereas angelica lactone may be an intermediate in the synthesis of pharmaceutical products. Hydrogenation may also form 1,5-pentanediol, n-pentanol, and valeric acid. This latter compound has application as plasticizer.

The reaction of levulinic acid with phenol may afford diphenolic levulinic acid, which may replace bisphenol-A (BPA) in polymers and other uses. BPA has been suspected to have mutagenic activity and its use in polymers that may have direct contact with food is being discontinued. Levulinic acid ketals may be produced upon the acid-catalyzed reaction with diols or triols, such as glycerol, obtained as a by-product of biodiesel production. The oxidative degradation of the keto group may afford succinic acid, which is a product with increasing applications.

Nitrogenated compounds can be also obtained from levulinic acid. The γ-amino levulinic acid may be synthesized in two steps from levulinic acid or levulinate esters. This compound may be useful in the synthesis of agrochemicals and used as pesticides [21]. N-Alkyl-pyrrolidones are also produced in two steps from levulinic acid or levulinate esters, being important intermediates in the pharmaceutical industry.

1.2 Global Market

Today, levulinic acid is industrially produced from biomass, being a renewable platform for the chemical industry. Table 1.2 shows the major producers and countries of origin. China appears as the major producer, but there are plants in the United States and Europe. The large number of companies indicates that levulinic acid is still produced on a relatively small scale, showing great potential to expand its use and, therefore, the capacity of the plants.

Although the production capacity is not always reported by the companies, Biofine Technology seems to be the largest producer of levulinic acid in the world, being able to use a variety of biomass feedstocks. GF Biochemicals, based in Italy, also appears as an important producer. The company has acquired smaller companies like Segetis and established a joint venture with the American Process Inc. to build an integrated biorefinery in the United States. More recently, it announced the formation of a joint venture with the Towell Engineering Group for the production and commercialization of levulinic acid.

Table 1.2 Main industrial producers of levulinic acid.

Company	Country
Biofine Technology LLC	United States
GF Biochemicals Co.	Italy
Bio-on	Italy
Langfang Hawk Technology & Development	China
Heroy Chemical Industry Co.	China
Parchem Fine & Specialty Chemicals	United States
Avantium Inc.	Holland
DSM	Holland
E. I. du Pont de Nemours and Company	United States
Hebei Langfang Triple Well Chemicals Co.	China
Apple Flavor & Fragrance Group Co.	China
China Shijiazhuang Pharmaceutical Group Co.	China
Hefei TNJ Chemical Industry Co.	China
Haihang Industry Co.	China
Tokyo Chemical Industry Co.	Japan

Bio-on is an Italian company that launched, in 2017, a project to produce levulinic acid together with the Sadam Group, which is also based in Italy and is known in the food and agroindustry sectors.

The Langfang Hawk Technology & Development Co. Ltd. is a Chinese company founded in 1992. Since 2002, it has been producing about 3 tons of levulinic acid per year from furfural. The Heroy Chemical Industry Co. produces levulinic acid in China since 1995. The other companies may produce levulinic acid as an intermediate for other processes or may simply commercialize it for resales. For instance, DSM announced a project to produce adipic acid from levulinic acid in 2016 and Avantium produces methyl levulinate in Holland.

The global market of levulinic acid was US$27.2 million in 2019, and it is expected to increase at a compound annual growth rate (CAGR) of 8.8% between 2020 and 2030 [22]. Global sales in 2030 are expected to reach US$60.2 million. The coronavirus pandemic may have changed this forecast, but levulinic acid is still a promising renewable raw material and the increasing demand will continue. In the year 2000, levulinic acid was commercialized with prices ranging from US$8.8–13.2 per kilogram. The market was relatively small, with total production around 0.5 ton per year. The market experienced a significant increase, and in 2013, the total production reached 2.6 ton, with prices ranging from US$5–8 per kilogram. North America is the world's largest consumer of levulinic acid and its derivatives, with Europe and Asia-Pacific coming next.

1.3 Direct Uses

The main direct use of levulinic acid is in cosmetics (Figure 1.5) to help prevent microbial growth, without altering the pH and the color of the beauty products. Sodium levulinate may be produced from the reaction of levulinic acid with basic solutions of sodium hydroxide or sodium carbonate, and it is also popular as a preservative in skin cosmetic formulations. According to the Voluntary Cosmetic Registration Program (VCRP) of the Food and Drug Administration (FDA) of the United States, 131 cosmetic formulations use levulinic acid, whereas sodium levulinate is present in 402 cosmetic preparations. The maximum concentration of LA is 4.5% in hair dyes, whereas sodium salt is present in up to 0.62% in mouthwashes and breath fresheners. Levulinic acid and sodium levulinate are used in baby products, such as lotion, oil, and cream formulations.

Levulinic acid and sodium levulinate may also find applications in the food industry, especially as preservatives in meats [23] and sausages [18]. They are effective in stunting bacterial growth. For instance, the addition of 1% or 2% of sodium levulinate in beef bologna and turkey roll, respectively, prevents the growth of *Listeria monocytogenes* upon refrigerated storage [24].

Levulinic acid has also been reported to enhance dermal penetration of drugs [25], whereas sodium levulinate was tested as a renewable deicing agent [26]. For instance, at −2.7 °C, solutions of sodium levulinate completely melted the ice.

Figure 1.5 Ingredients of a commercial baby moistening cosmetic, including levulinic acid in its formulation (https://world.openbeautyfacts.org/product/4311596611461/baby-pflegecreme-blutezeit). Source: Open Beauty Facts.

Decarboxylation of sodium levulinate yields methyl ethyl ketone (MEK) (Scheme 1.6), an important solvent in the chemical industry [27]. The process involves an electrochemical cell that is fed with sodium levulinate, sodium hydroxide, water or methanol, and hydrogen. The decarboxylation of sodium levulinate generates free radicals that may interact with hydrogen gas to yield MEK and octanedione.

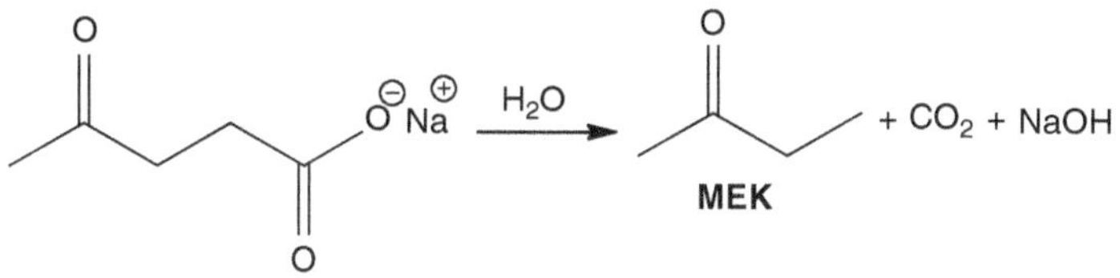

Scheme 1.6 Decarboxylation of sodium levulinate to methyl ethyl ketone (MEK).

Calcium levulinate is another important salt of levulinic acid. It has been used as a dietary supplement of calcium for more than 80 years [28]. It may be presented as pills, capsules, or injections in the pharmaceutical industry (Figure 1.6) to serve as a food nutrition enhancer that improves bone formation and muscle excitability. Calcium levulinate has also been used in beverages [29] and vitamins [30], as a source of calcium.

Calcium levulinate is normally precipitated in the form of dihydrate of molecular formula $Ca(C_5H_7O_3)_2(H_2O)_2$, which forms a polymeric structure with the

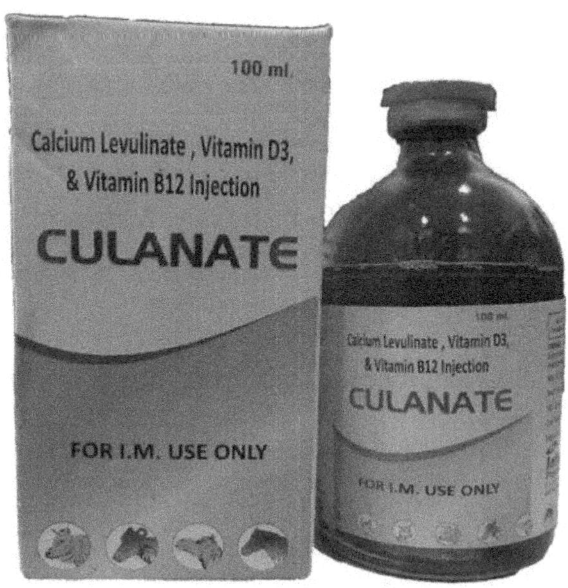

Figure 1.6 Veterinary supplement containing calcium levulinate (https://shopsi.martagaska.com/category?name=5000%20iu%20v%20ug). Source: shopsi.martagaska.com.

Figure 1.7 Three-dimensional structure of calcium levulinate, highlighting the unit cell and the hydrogen bonds (dotted lines) [31]. Source: Amarasekara et al. [31]/International Union of Crystallography/CC BY 2.0 UK.

calcium cations being octacoordinated, with two aqua ligands and six oxygen atoms from the carboxylate groups [31]. The three-dimensional arrangement presents interchain hydrogen bonds, involving the water molecules of one chain and the carboxylate groups of the other chain (Figure 1.7).

1.4 Toxicity of Levulinic Acid and Inorganic Levulinates

Levulinic acid is listed by the FDA of the United States as a permitted synthetic flavoring substance and adjuvant in food for human consumption [32]. The acute dermal toxicity (LD_{50}) of levulinic acid in rabbits is greater than 5000 mg kg^{-1}, whereas the acute oral LD_{50} in rats is reported to be 1850 mg kg^{-1} [33]. Levulinic acid did not show any short-term toxicity in rats and adult male humans [34].

Table 1.3 Typical concentration of levulinic acid and sodium levulinate in cosmetic and personal care formulations.

Product	Levulinic Acid (%)	Sodium Levulinate (%)
Shampoos	0.48	—
Hair dyes and color	4.5	—
Mouth washes and breath fresheners	0.35	0.62
Baby lotions	—	0.35
Eye shadows	—	0.57
Body and hand products	—	0.002

No significant abnormality was observed in the necropsy of the rats, whereas the humans did not show any immediate hematological effect related to the tests.

Nondiluted solutions of levulinic acid cause moderate skin irritation in rabbits, but no irritation in human skin was observed with diluted solutions [32]. Levulinic acid is not suspected to have mutagenic activity but may cause severe eye irritation. No respiratory or skin sensitization is known, as well as carcinogenicity and reproductive toxicity. Thus, levulinic acid and sodium levulinate may be considered safe and nontoxic for applications in small concentrations in cosmetics and food preservation. Table 1.3 shows some typical concentrations of levulinic acid and sodium levulinate in cosmetic formulations.

Calcium levulinate is commercially available as an oral or intravenous calcium supplement. It is considered safe for application in humans, because of its high solubility in water, good calcium content (13.1 wt%), and almost neutral pH of the solutions, between 7 and 8. Nevertheless, there are concerns about the metabolism of levulinate in the organism.

Studies in rats have indicated that levulinate is converted to 4-hydroxy-pentanoate [35], a potential new drug of abuse. In addition, levulinate metabolism leads to a substantial accumulation of levulinyl-CoA, 4-hydroxypentanoyl-CoA, and 4-phosphopentanoyl-CoA in the brain and in the liver, especially in the presence of ethanol. Therefore, the concomitant ingestion of calcium levulinate and alcoholic beverages may lead to an increase of the concentration of 4-hydroxy-pentanoate in the body, which raises important health public concerns.

1.5 Concluding Remarks

Levulinic acid is a potential new renewable platform chemical, obtained from the acid hydrolysis of pentoses and hexoses. It has been pointed out by the US

Department of Energy (DOE) [36] among the 12 most important building block molecules (Figure 1.8) that could be obtained from biomass, with the potential to be transformed into value-added products. The list has been revisited [37, 38] since then, but levulinic acid has remained among the most promising bio-based chemical platform.

The global production of levulinic acid is still low, but the forecast indicates a significant increase in demand in the forthcoming years. There are major producers in the United States, Europe, and Asia and, as the production increases, the price will decrease making levulinic acid an important commodity for the production of fuels and chemicals.

Levulinic acid and sodium levulinate are mostly used as preservatives in cosmetics and beverages. Calcium levulinate finds applications in pharmaceutic formulations and beverages, as a supplementary source of calcium for the organism. These uses, however, cannot justify the increase in the world's production of levulinic acid.

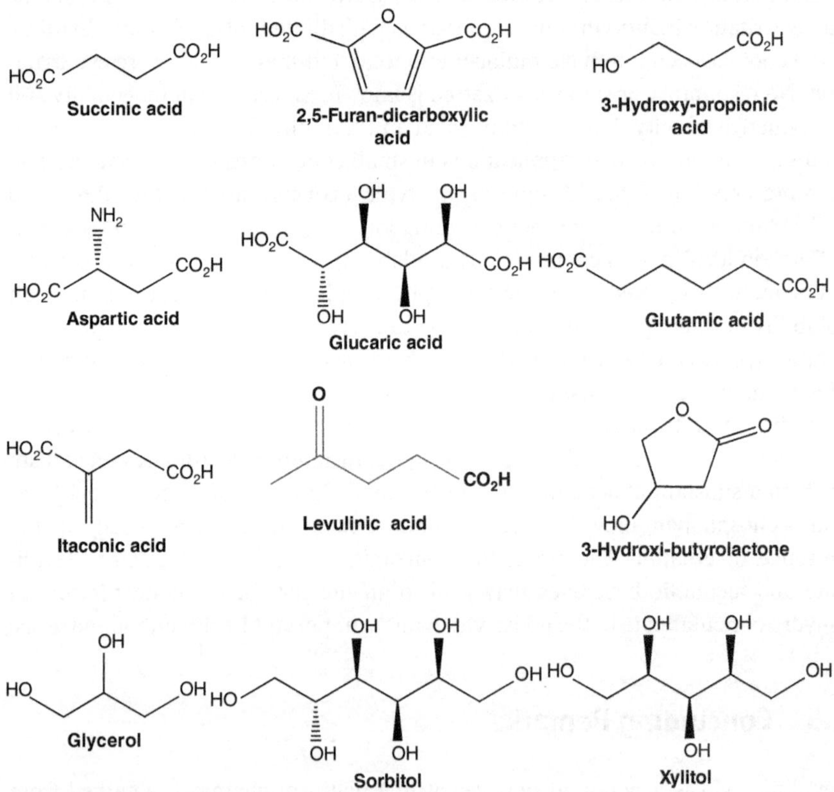

Figure 1.8 Renewable chemical platforms listed by the US DOE.

The next chapters will focus on the main current technologies of levulinic acid production from biomass, as well as the main derivatives of potential commercial interest. Chapter 2 describes the synthetic routes and technologies for the production of levulinic acid. The emphasis is on the use of different biomasses as feedstocks. The chapter also highlights the most important commercial initiatives for the production of levulinic acid.

Chapter 3 is dedicated to levulinates, mostly the ester derivatives, which can be obtained from esterification of levulinic acid or by direct conversion of the biomass in alcoholic medium. Organic levulinates not only find applications as flavoring agents but also have potential as fuel additives.

Hydrogenation of levulinic acid is the topic of Chapter 4. The main focus is the production of GVL, which can be directly used as solvent and fuel additive. Nevertheless, it may be used as a starting material for the production of hydrocarbons, through complex chemical reactions.

Chapter 5 highlights the reactions in the carbonyl group of LA. There are many important derivatives with potential commercial applications. Finally, Chapter 6 remarks some potential transformation of levulinic acid in the context of a biorefinery.

References

1. Malaguti, F. (1835). Action des Acides Étendus sur le Sucre. *Compt. Rendus.* 1: 59–60.
2. Mulder, G.J. (1840). Untersuchungen uber die humussubstanzen. *J. Prakt. Chem* 21: 203–240.
3. Freiherrn, A., Grote, V., and Tollens, B. (1875). Untershuchugen über Kohlenhydrate. I. Ueber die bein Einwirkung von Schwefelsäure auf Zucker entstehende Säure (Levulinsäure). *Justus Liebigs Ann. der Chem.* 175: 181–204.
4. Frost, T.R. and Kurth, E.F. (1951). Levulinic acid from wood cellulose. *Tappi* 34: 80–86.
5. Wendell, W.M. (1942). Preparation of levulinic acid. US Pat 2270328; A. E. Staley Co., issued 20 January 1942.
6. Brownlee, H.J. (1936). Process for production of furfural. US Pat 2140572; Quaker Oats Co., issued 24 June 1936.
7. Dashtban, M., Gilbert, A., and Fatehi, P. (2012). Production of furfural: overview and challenges. *J. Sci. Technol. Forest Prod. Proc.* 2: 44–53.
8. Biomass-Based Chemicals (2022). http://www.furan.com/furfuryl_alcohol_historical_overview.html.
9. Swadesh, S. (1952). Catalytic production of furfuryl alcohol and catalyst therefor. US Pat 2, 754,304; Quaker Oats Co., issued 27 May 1952.

10 Hart, L.R. and Kenneth, R. (1956). Process for the manufacture of levulinic acid esters. US Pat. 2763665; Howard of Ilford Limited, issued 18 September 1956.

11 Quaker Oats Company (1959). *Ind. Eng. Chem.* 51 (1): 54A–55A.

12 Zanghellini, A.L. (2012). Fermentation route for the production of levulinic acid, levulinate esters, valerolactone and derivatives thereof. WO Pat 030860 Al, Arzeda Corp., issued 28 September 2017.

13 Farnleitner, L., Stuckler, H., Kaiser, H., and Kloimstein, E. (1990). Verfahren zur Herstellung lagerstabiler Lävulinsäure. EP 0401532 A1; Chemie Linz Gesellschaft m.b.H., issued 12 December 1990.

14 Wiebusch, K. (1963). Production of maleic acid and maleic anhydride. US Pat. 3086026; BASF Co., issued 16 April 1963.

15 Skinner, W.A. and Tieszen, D. (1961). Production of maleic acid by oxidizing butenes. *Ind. Eng. Chem.* 53: 557–558.

16 Wojcieszak, R., Santarelli, F., Paul, S. et al. (2015). Recent developments in maleic acid synthesis from bio-based chemicals. *Sustainable Chem. Process* 3: 9. https://doi.org/10.1186/s40508-015-0034-5.

17 Wang, Z.W., Lei, T.Z., Liu, L. et al. (2012). Performance investigations of a diesel engine using ethyl levulinate-diesel blends. *BioResources* 7: 5972–5982.

18 Vasavada, M., Carpenter, C.E., and Cornforth, D. (2003). Sodium levulinate and sodium lactate effects on microbial growth and stability of fresh pork and turkey sausage. *J. Muscle Foods* 14: 119–129.

19 Bereczky, A., Lukács, K., Farkas, M., and Dóbé, S. (2014). Effect of γ-valerolactone blending on engine performance, combustion characteristics and exhaust emissions in a diesel engine. *Nat. Resourc.* 5: 177–191.

20 Aycock, D.F. (2006). Solvent applications of 2-methyltetrahydrofuran in organometallic and biphasic reactions. *Org. Proc. Res. Dev.* 11: 156–159. https://doi.org/10.1021/op060155c.

21 Sasikala, C., Ramana, C.V., and Rao, P.R. (1994). 5-Aminolevulinic acid: a potential herbicide/insecticide from microorganisms. *Biotechnol. Prog.* 10: 451–459.

22 Report Levulinic Acid Market. https://www.psmarketresearch.com/market-analysis/levulinic-acid-market.

23 Carpenter, C.E., Smith, J.V., and Broadbent, J.R. (2011). Efficacy of washing meat surfaces with 2% levulinic, acetic, or lactic acid for pathogen decontamination and residual growth inhibition. *Meat Sci.* 88: 256–260.

24 Thompson, R.L., Carpenter, C.E., Martini, S., and Broadbent, J.R. (2008). Control of *Listeria monocytogenes* in ready-to-eat meats containing sodium levulinate, sodium lactate, or a combination of sodium lactate and sodium diacetate. *J. Food. Sci.* 73: M239–M244.

25 Taghizadeh, S.M., Moghimi-Ardakani, A., and Mohamadnia, F. (2015). A statistical experimental design approach to evaluate the influence of various penetration enhancers on transdermal drug delivery of buprenorphine. *J. Adv. Res.* 6: 155–162.

26 Ganjyal, G., Fang, Q., and Hann, M.A. (2007). Freezing points and small-scale deicing tests for salts of levulinic acid made from grain sorghum. *Bioresour. Tech.* 98: 2814–2818.

27 Karanjikar, M. and Bhavaraju, S. (2014). Decarboxylation of levulinic acid to ketone solvents. *US 8853463 B2*; Ceramatec Inc., issued 7 October 2014.

28 Proskouriakoff, A. (1933). Some salts of levulinic acid. *J. Am. Chem. Soc.* 55: 2132–2134.

29 Livisay, S. and Lavoie, J. (2003). Calcium-fortified, grape-based products and methods for making them. US 0203073A1; Welch Foods, Inc., issued 13 November 2002.

30 Paradissis, G.N., Levinson R.S., Heeter, G., Cuca, R.C., and Vanek, P.P. (1999). Multi-vitamin and mineral supplements for women. US 5869084A; Kv Pharmaceuticals Co., issued 9 February 1999.

31 Amarasekara, A.S., Sterling-Wells, D.T., Ordonez, C. et al. (2015). Crystal structure of a polymeric calcium levulinate dihydrate: *catena*-poly[[diaquacalcium]-bis(μ_2-4-oxobutanoato)]. *Acta Cryst.* E71: 494–497.

32 U.S. Food and Drug Administration, U.S.Department of Health & Human services. https://www.accessdata.fda.gov/scripts/cdrh/cfdocs/cfcfr/cfrsearch.cfm?fr=172.515.

33 Anonymous (1979). Laevulinic acid. *Food Cosmet. Toxicol.* 17: 847–848.

34 Tischer, R.G., Fellers, C.R., and Doyle, B.J. (1942). The nontoxicity of levulinic acid. *J. Am Pharm Assoc.* 31: 217–220.

35 Harris, S.R., Zhang, G.F., Sadhukhan, S. et al. (2011). Metabolism of levulinate in perfused rat livers and live rats: conversion to the drug of abuse 4-hydroxypentanoate. *J. Biol. Chem.* 286: 5895–5904.

36 Werpy, T., Petersen, G., Aden, A. et al. (2004). Top value-added chemicals from biomass. https://www.nrel.gov/docs/fy04osti/35523.pdf (accessed 7 August 2022).

37 Bozell, J.J. and Petersen, G.R. (2010). Technology development for the production of biobased products from biorefinery carbohydrates—the US Department of Energy's "Top 10" revisited. *Green Chem.* 12: 539–554.

38 Takkellapati, S., Li, T., and Gonzalez, M.A. (2018). An overview of biorefinery derived platform chemicals from a cellulose and hemicellulose biorefinery. *Clean Technol. Environ. Policy* 20: 1615–1630.

25 Tagliabue, S.M., [illegible] (2019). [illegible]
[illegible]
[illegible]
[illegible]
26 Cappai, G., [illegible] and Chen, M.A. (2017). Heavy metals and [illegible]
[illegible]
[illegible]
27 [illegible] (2019). [illegible] Ozone [illegible] 2019.
28 [illegible]
29 Leroy, [illegible] (2017). [illegible]
[illegible] 15 November 2019.
30 [illegible]
[illegible]
[illegible]
31 [illegible]
[illegible]
[illegible]
32 [illegible] (2011). [illegible]
[illegible]
[illegible]
33 Anonymous (1978). [illegible]
34 [illegible]
[illegible]
35 [illegible]
[illegible]
36 [illegible] (2004). [illegible]
[illegible]
2007.
37 [illegible]
[illegible]
[illegible]
38 [illegible]
[illegible]
[illegible]

2

Production and Technological Routes

2.1 Production and Technological Routes from Biomass

Levulinic acid (LA) has been synthesized since 1870 through various methods, such as hydrolysis of acetyl succinate ester, acid hydrolysis of furfuryl alcohol, oxidation of ketones with ozone, carbonylation of ketones, and alkylation of nitroalkanes. However, all the mentioned processes involve large number of steps, consume expensive reagents, and generate wastes that may lead to environmental problems. An alternative route for the synthesis of LA is agriculture residues rich in lignocellulosic biomass, considered an abundant, cheap, and renewable resource [1–6].

Fossil sources, including petroleum, coal, and natural gas, are causing environmental concerns at the beginning of the twentieth century due to climate changes and the strong volatility of their prices. Therefore, the industrial sector searches for alternative raw materials that could replace oil-based products. In addition, with the projection of the world's population reaching 10.5 billion people by 2050, there will be strong demands for energy, food, chemicals, and services [7].

The production of LA from renewable raw materials is a highly attractive option. However, there are several challenges to be faced considering the large-scale production, such as availability of raw materials, adaptation of technologies, organization of the production and supply chains, logistics, seasonality, and influence of climatic cycles, among other factors. Certainly, it requires an intense process of experimentation [8].

Starting from lignocellulosic material, several chemicals can be obtained owing to multiple functionalities (Scheme 2.1). Among them, LA has been listed as one with the highest added-value by the US Department of Energy (DOE) and has also been identified as a promising sustainable material for the synthesis of important chemicals of great industrial application, such as corrosion inhibitors, adsorbents,

Levulinic Acid: A Sustainable Platform Chemical for Value-Added Products, First Edition.
Claudio J.A. Mota, Ana Lúcia de Lima, Daniella R. Fernandes, and Bianca P. Pinto.

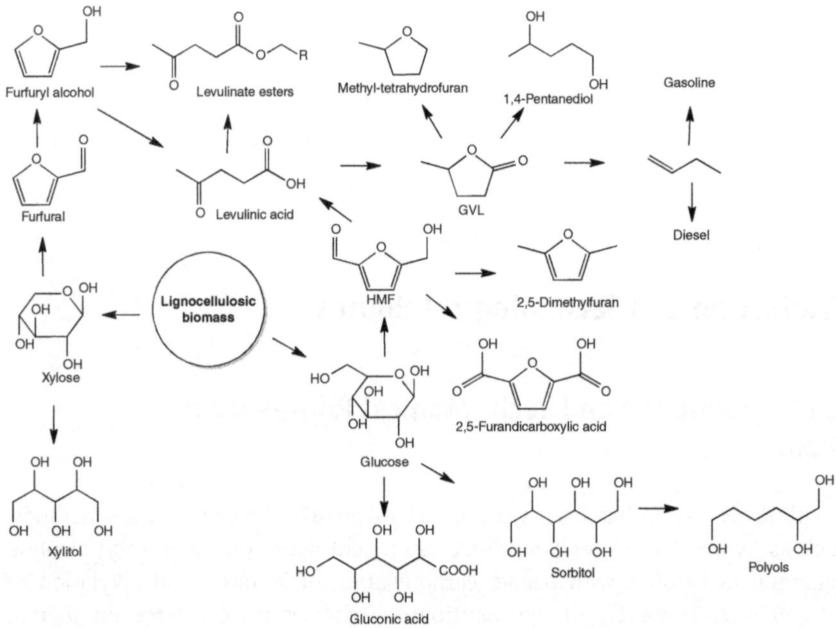

Scheme 2.1 Reaction pathways for renewable fuels and chemicals from lignocellulosic biomass. Source: Wettstein et al. [9]/with permission of Elsevier.

coatings, polymers, antifreeze agents, electronics, plasticizers, flavoring agents, antifouling compounds, herbicides, solvents, among others [10–12].

The use of lignocellulosic biomass for the production of LA is extremely attractive due to its renewability, abundance, low cost, and neutral carbon balance. In addition, lignocellulosic materials do not directly influence food production, but compete with other applications, such as destination for animal feed, industrial products, and their use for energy production [9, 13].

The supply of lignocellulosic residues around the world corresponds to approximately 2.9 billion tons produced from cereals, 3 billion tons from seeds, and 0.5 billion ton from other types of crops, in addition to 40 million tons of inedible plant materials, which are mostly discarded as waste. In this scenario, Brazil stands out worldwide as a promising land for agriculture and agribusiness, with a large arable area of grains, cereals, fruits, and especially the cultivation of sugar cane [14, 15].

Lignocellulosic materials are complex blends of natural polymers, consisting of cellulose (35–50%), hemicelluloses (25–30%), and lignin (15–30%) tightly bound by physical and chemical interactions. The variation of composition is dependent on the type of biomass. Sugarcane bagasse, corn straw, hardwood, coniferous wood,

Table 2.1 Typical general composition of different biomasses.

Lignocellulosic materials	Cellulose (%)	Hemicellulose (%)	Lignin (%)
Sugarcane bagasse	40–45	30–35	20–30
Sweet sorghum bagasse	45	27	21
Wheat straw	33-40	20-25	15–20
Barley straw	35	29	13
Soybean straw	34	16	22
Sorghum straw	27	33	10
Rice straw	38	32	12
Corn stover	38	23	20
Corn cob	41	31	12
Poplar	44	20	29
Rice husk	40	16	26
Pine	42	21	30
Switchgrass	31	24	18
Coastal Bermuda grass	30	29	23
Napier grass	47	31	22
Elephant grass	36	24	28
Bamboo	45	24	20
Softwood spruce	25	10	35
Pinewood	38	24	34
Eucalyptus	45	29	26
Waste papers	65	13	1

Source: Adapted from [16, 17].

cellulosic residues, and herbaceous biomass are some examples of lignocellulosic materials. Table 2.1 shows the typical general composition of different biomasses [18, 19].

Cellulose, the main component of lignocellulosic material, is a polysaccharide formed by glucose molecules joined by β-1,4-glycosidic bonds. Cellulose is the most abundant renewable resource found in nature. Its chains are stabilized and linked together by intra- and intermolecular hydrogen bonds, forming elementary fibrils. These fibrils are joined together through hemicellulose, which is an amorphous polysaccharide composed of several sugar molecules with five and six carbon atoms, such as arabinose, galactose, glucose, mannose, and xylose, among others, in addition to deoxysugars and uronic acids [20, 21].

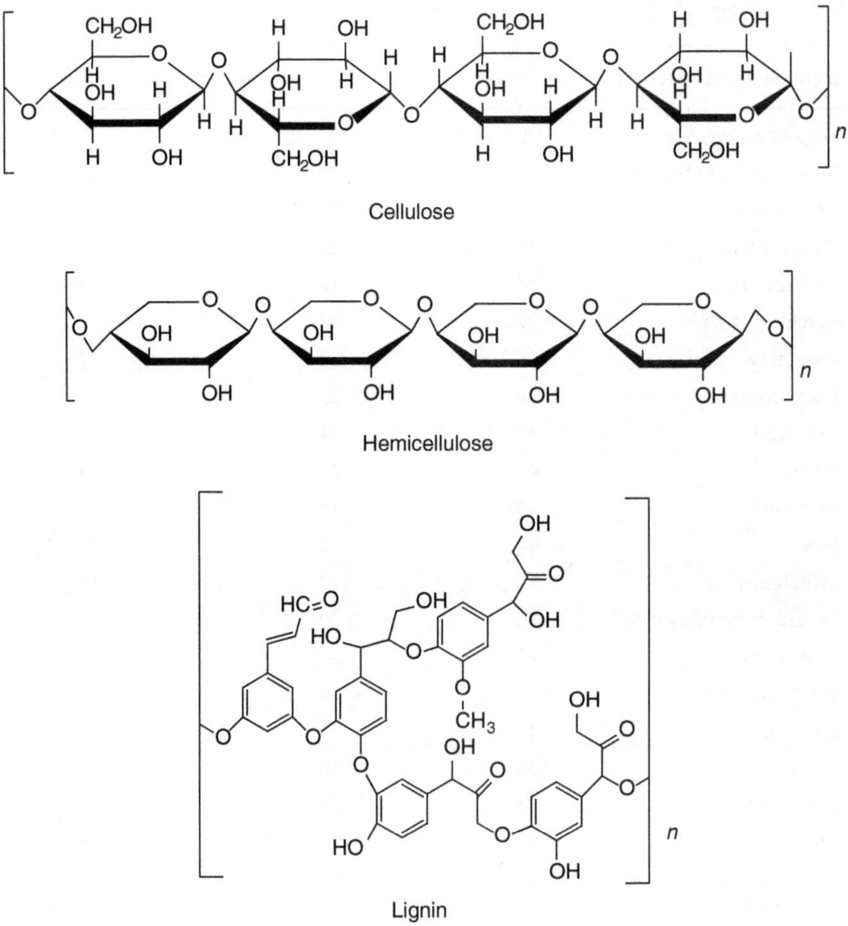

Figure 2.1 Schematic structure of cellulose, hemicellulose, and lignin.

Cellulose and hemicellulose chains are enveloped by lignin, a complex aromatic polymer formed by three basic phenolic structures: p-coumaryl alcohol, coniferyl alcohol, and synapyl alcohol, as well as its derivatives. Lignin is undesirable in the LA production processes, as it can lead to unfavorable results. In addition, residual solids associated with lignin and humins tend to clog the reactors. Figure 2.1 shows the structure of cellulose, hemicellulose, and lignin [20, 21].

The chemical transformation of cellulose poses many challenges due to the problems associated with its structural protection mechanisms, which make this polymer highly resistant to chemical and biological transformations. Consequently, the hydrolysis of cellulose requires extreme conditions, such as the use of strong acids and high temperatures [22].

2.2 Pretreatment of Lignocellulosic Biomass

Pretreatment is an obligatory step in the chemical conversion of lignocellulosic biomass to LA due to the lignin component that envelopes the cellulose and hemicellulose units. Therefore, it is of prime importance to deconstruct the heterogeneous lignocellulose material and to determine the conversion rates. The pretreatment access the basic constituents and leads to a decrease in the particle size, physical opening, and separation of cellulose and hemicellulose, as shown in Figure 2.2 [22–24].

The ideal pretreatment produces mono and oligosaccharides that would be easily hydrolyzed, but without forming degradation products and fermentation inhibitors. In general, the pretreatment methods fall into four different categories, including physical, chemical, physicochemical, and biological (Figure 2.3). In many cases, more than one type of pretreatment may be used [25–27].

2.2.1 Physical Pretreatment

The physical pretreatment consists of grossly reducing the particle size of the lignocellulosic material through crushing, chipping, and grinding. These treat ments increase the available surface area, reduce the degree of polymerization, and considerably reduce the crystallinity of the biomass. In general, mechanical pretreatment is efficient in removing lignin from the cell wall. There are three

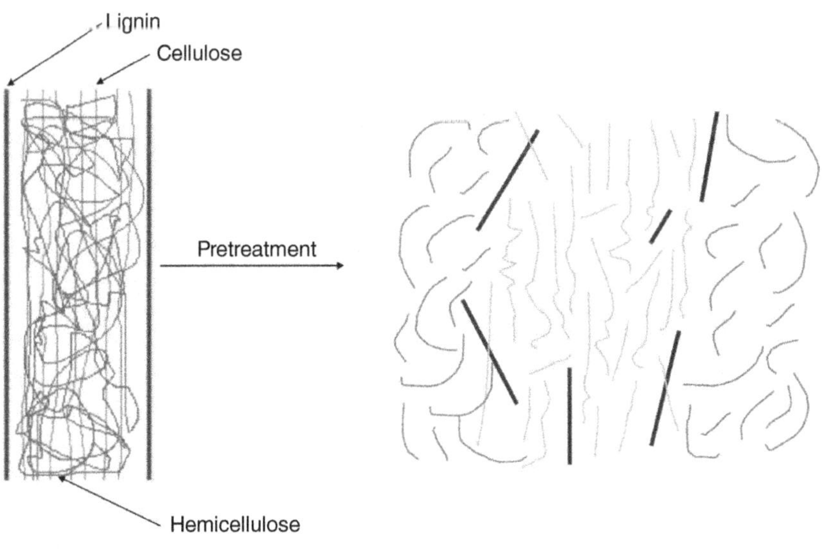

Figure 2.2 Structural alterations of lignocellulosic biomass after pretreatment.

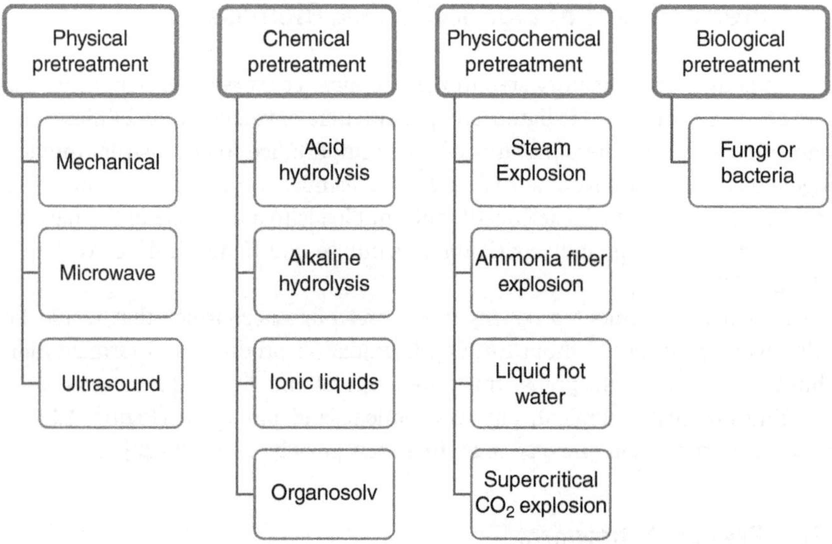

Figure 2.3 Main types of lignocellulosic biomass pretreatments.

main types of physical pretreatments, namely mechanical, microwave, and ultrasound [28, 29].

2.2.1.1 Mechanical

Mechanical pretreatment methods include techniques such as milling, grinding, chipping, or extrusion. The particle size can be reduced to 0.2 mm, depending on the type of grinding method adopted, processing time and, also, the type of biomass used. The great advantage is the nonformation of fermentation inhibitors. Additionally, there is no need to wash the pretreated material. However, the milling pretreatment requires high-energy input and capital expenditure associated with the mechanical equipment. Since the milling pretreatment does not result in any toxic products or inhibitor compounds, it is a preferred preliminary pretreatment for a wide variety of lignocellulosic raw materials [30, 31].

In extrusion, the raw materials are passed through a cylinder at a high temperature (above 300 °C), which allows the disruption of the lignocellulosic structure due to the combination of high temperature and shear forces, caused by the rotating blades present in the extrusion machines. However, there is high-energy consumption, which makes this pretreatment process less attractive. The combination of physical and chemical treatments may reduce the energy consumption [32, 33].

2.2.1.2 Microwave

Microwave irradiation is an unconventional heating method that results in disruption of the lignocellulosic structure. It has several advantages, such as easy operation, energy efficiency, minimum formation of inhibitors, and high-heating capacity within a short period. Microwave pretreatments can be carried out under atmospheric conditions or at high pressures. In this later case, it is necessary to use closed reactors operating at temperatures in the range of 150–250 °C. Microwave pretreatments are often used in combination with chemical pretreatments, such as the use of acids, and bases or ionic liquids. Low temperatures are generally preferred to avoid hemicellulose degradation and to improve the pentose yield [34–36].

Microwave pretreatment may increase the accessibility of the cellulosic units for subsequent enzymatic hydrolysis. For example, when irradiated with microwaves, rice straw, and bagasse show increased accessibility. Although being a rapid pretreatment process, microwave irradiation requires high-energy inputs, which should be considered upon scaling up the process [37].

2.2.1.3 Ultrasound

Cavitation generates shear forces that cleave the structure of the lignocellulosic materials and promote the extraction of the desired units. The choice of solvent is important to determine optimal pretreatment conditions; other factors such as temperature, ultrasound frequency, and duration are also relevant. It is an energy-intensive process, and it is necessary to optimize the process parameters for large-scale applications [38, 39].

Studies have reported a decrease in particle size and/or increase in surface area of sonicated substrates. For example, the particle size of cassava paste can be decreased by 40-folds after ultrasound irradiation for 10–40 seconds at $8\,W\,ml^{-1}$, as the result of the breakdown of the lignocellulosic material. As consequence, it was possible to obtain a reducing sugar yield of 22/100 g of the sample [40].

2.2.2 Chemical Pretreatment

This pretreatment route uses acids, alkalis, organic solvents, and ionic liquids for the deconstruction of the biomass structure. The most common chemical pretreatments are acid and alkali hydrolysis [25].

2.2.2.1 Acid Hydrolysis

Acid hydrolysis is one of the most extensively employed biomass pretreatment. One of the major problems of acid hydrolysis is the formation of fermentation inhibitors, which occurs because of severe reaction conditions, such as strong acidity and high temperature. To try to minimize these problems, dilute acids and mild temperature conditions, up to 120 °C, are used [25].

Inhibitors are formed from the degradation of sugars or lignin, generating products that can be classified into three groups: weak acids, furan derivatives, and phenolic compounds. When this occurs, it is necessary to detoxify the biomass to reduce the concentration of inhibitors. In addition, it is necessary an additional step of neutralization to eliminate the formed inhibitors, which increases the process costs and generates liquid effluents [41]. The use of ion exchange resins and activated carbon may overcome the effects of inhibitor formation. They convert the inhibitor compounds into inert substances or reduce their concentration. However, the selection of the raw materials that produce fewer inhibitors is the simplest approach to alleviate the problem [42].

In the acid pretreatment, hydronium ions from acid dissociation cause the breakdown of the long cellulose and hemicellulose chains yielding mono or oligosaccharide sugars. Inorganic acids such as sulfuric, hydrochloric, nitric, phosphoric, as well as organic acids such as formic and oxalic, are widely used. It is important to note that most concentrated acid solutions are toxic and highly corrosive, resulting in high operating and maintenance costs. Of all chemical pretreatments, dilute sulfuric acid, within 0.5–1.5%, at temperatures above 160 °C, has been the preferred process conditions for industrial applications, affording sugar yields within 75–90% [43–45].

2.2.2.2 Alkaline Hydrolysis

Alkaline pretreatment is a widely studied chemical method, based on the solubilization of lignin in the alkaline solution. The pretreatment is more effective with agricultural residues of low-lignin content but becomes less effective as the lignin content of the biomass increases [46].

During alkaline pretreatment, there occurs saponification leading to the cleavage of the intermolecular bonds between hemicelluloses and lignin, resulting in the solubilization of the lignin and hemicellulose fragments. There is also a change in the lignocellulosic structure through swelling of the cellulose, which leads to a reduction of crystallinity and the degree of polymerization, increasing the surface area. Furthermore, the removal of acetyl groups and uronic acid substitutions in hemicelluloses, during alkaline pretreatment, also increases the accessibility of the carbohydrate chains toward hydrolysis [47].

Sodium, potassium, and calcium hydroxides, hydrazine, and anhydrous ammonia are some of the most used alkaline pretreatment media. For example, the pretreatment of rice straw with 1% aqueous NaOH at room temperature for three hours significantly reduces the hemicellulose and lignin contents, while the cellulose content remains unchanged. Similarly, the pretreatment of sugarcane bagasse with 1% aqueous NaOH at 100 °C for three hours caused about 86% lignin removal [48–51].

2.2.2.3 Ionic Liquids

Ionic liquids are salts consisting of an inorganic anion and an organic cation moiety, which can act as solvents or catalysts. The vast majority of ionic liquids are recoverable and reusable. They feature thermal and chemical stability, being nonvolatile and usually nontoxic upon handling. Ionic liquids can be easily separated from the reaction medium and allow the reactions to be carried out at low temperatures. Imidazolium salts are one of the most used ionic liquids, although efficient, imidazolium cations are expensive and, therefore, limited in their industrial deployment at a large scale [52, 53].

Ionic liquids have potential application for the pretreatment of lignocellulosic biomass because they can be employed at low temperatures. During the pretreatment, the cations and anions play a significant role in the solubilization of cellulose and lignin. The number of potential ionic liquids that can be used in biomass pretreatment is large, because of the possibility to make different combinations. However, ionic liquids are considered toxic to microorganisms and enzymes and sensitive to impurities. Thus, they still present obstacles to being used on an industrial-scale biomass pretreatment process. Table 2.2 shows some types of ionic liquids applied for the pretreatment of different biomasses [74, 75].

The recyclability and reuse of ionic liquids are another point of attention. The ionic liquid [C_2MIM][OAc] was reused seven times in the wood residue solubility step, contributing mainly to the breakdown of the lignocellulosic structure and partial removal of lignin. There are reports that the ionic liquid ([Emim][OAc]) was successfully recycled more than 20 times during the pretreatment of rice straw, reaching over 90% glucose conversion in the first cycle and 78% in the twentieth cycle. The recycling and reuse of ionic liquids are essential steps for its large-scale application, since ionic liquid are expensive [76, 77].

2.2.2.4 Organosolv

Organosolv is a type of pretreatment that uses organic solvents at temperatures ranging from 100 to 250 °C. Several types of organic solvents are used such as alcohols, phenols, esters, acetone, formaldehyde, dioxane, and amines with and without the use of catalysts. Due to their low boiling point, low cost, and easy recovery, methanol, and ethanol are the most used solvents for this process [78]. Organosolv pretreatment is effective in overcoming biomass recalcitrance, but it is also an efficient method of lignin fractionation [79].

During organosolv pretreatments, the lignin is solubilized in the organic solvent and can be recovered upon removing the organic solvent from the liquid phase or diluting it in water. Thus, the structural characteristics of the lignin obtained

Table 2.2 Some common ionic liquids applied in biomass pretreatment.

Biomass	Ionic Liquids	References
Cotton stalks	$HBIMHSO_4$	[54]
Corn stalks	$[Bmim]BF_4$	[55]
Wheat straw	[Emim][DEP]	[56]
Sugarcane bagasse	$[H_3N(CH_2)_2OH][OAc]$	[57]
Corn stalk	[Emim]Ac	[58]
Bamboo	EmimAc; BmimAc; AmimAc	[59]
Sugarcane bagasse	$[Me(NH_2)(CH_2)_2OH]$ [OAc]	[60]
Corn stalk	$[Bmim]BF_4$	[55]
Rice straw	[HMMorph][Cl]	[61]
Corn stalk	$([C_4C_1im]Cl)$	[62]
Eucalyptus sawdust	[TBA][OH]	[63]
Oil palm	([BMIM]Cl)	[64]
Spruce and oak sawdusts	[Emim][OAc]	[52]
Corn stover	[Ch][Lys]; $[C_2C_1Im][OAc]$	[65]
Bagasse powder	[Emim][OAc]	[66]
Triticale straw	([Emim]OAc)	[67]
Wheat straw	[Emim]OAc	[68]
Poplar	(EMIMOAc)	[69]
Cashew apple bagasse	(2-HEAA)	[70]
Fiber sludge	([AMIM]Cl); ([BMIM]Cl)	[71]
Agave bagasse	$([C_2mim][OAc])$	[72]
Rice straw	[EMIM][OAc]	[73]

through organosolv pretreatment strongly depend on the selection of the organic solvent used and the pretreatment conditions [79].

Most organic solvents are expensive and need to be recovered. In addition, this process may be energy-intensive, as distillation is the main process of solvent recovery. Another point of concern is that organic solvents are often volatile and flammable; thus, the pretreatment must be carried out under controlled conditions [80].

2.2.3 Physicochemical Pretreatment

The physicochemical pretreatment includes the vast majority of pretreatment technologies. Chemical agents such as acids, alkalis, or water along with physical processes like high temperatures and pressures are often used. These methods increase the surface area of the biomass material also hydrolyzing it partially or completely. The main processes are steam explosion, hot liquid water, ammonia fiber expansion, and supercritical CO_2 explosion [81, 82].

2.2.3.1 Steam Explosion (SE)

SE is a pretreatment technique that involves the exposure of the biomass to steam at high temperatures (150–250 °C) and pressures (20–50 bar) within short residence times (2–60 minutes), followed by rapid decompression. The process breaks the structure of the fibrous lignocellulosic material that is collected into solid pieces and as a liquid residue [83, 84].

The process occurs in two steps, the first being called autohydrolysis and the second decompression. In the first step, the steam condenses and permeates the lignocellulosic structure, removing the acetyl and uronic groups present in the hemicellulose chains. This leads to the formation of acetic and uronic acids, respectively, which act in the hydrolysis of the hemicellulose fraction. In the decompression step, the moisture condensed on the fibers evaporates again and promotes the mechanical breakdown of the lignocellulosic structure [84, 85].

The solid fragments possess a modified chemical structure, showing greater surface area. The liquid residue contains sugars, arisen from the partial hydrolysis of hemicellulose and cellulose, as well as soluble lignin fragments. The efficiency and selectivity of this pretreatment are highly dependent on the raw material and applied conditions [86].

After the steam explosion, the biomass becomes more susceptible to the action of cellulolytic enzymes. Several works show the efficiency of this pretreatment in improving the enzymatic digestibility of different lignocellulosic biomasses, such as sugarcane bagasse, eucalyptus residue, corn fiber, and wheat straw, among others [85–87].

Due to the high temperatures involved in the SE pretreatment, a major disadvantage is the formation of inhibitory compounds for fermentation processes, such as acetic acid, furfural, and hydroxymethyl furfural (HMF). In addition, it is necessary to wash the biomass to remove the toxic products. The main advantages are the good yield of sugars, no need for prior biomass grinding, low-energy consumption, and no cost for treatment of the effluents [88].

2.2.3.2 Liquid Hot Water (LHW)

LHW is a technique similar to steam explosion, which uses hot water or saturated steam to hydrolyze and dissolve hemicellulose fragments, while partially solubilizing the lignin with almost no influence on the cellulose fibers. Due to the lower temperatures employed, the formation of inhibitors is attenuated. The operating conditions depend on the biomass used. Generally, temperatures above 130 °C are required as no acid or base catalysts are employed. In addition, this pretreatment has lower costs in terms of capital investments. The LHW technique is widely used in the paper industry [89–92].

LHW is considered a simple and efficient method for separating recalcitrant compounds from biomass. Optionally, acids can be used to improve the performance. During LHW pretreatment, most hemicellulose fractions are separated and hydrolyzed to monomers or oligomeric sugars, which can be used as substrates to produce various compounds. Furthermore, the pH must be controlled between 4 and 7 to avoid the production of fermentation inhibitors. Higher pH values are beneficial for the formation of xylo-oligosaccharides, minimizing the formation of xylose and furfural derivatives [93, 94].

2.2.3.3 Ammonia Fiber Expansion (AFEX)

In this process, liquid ammonia is added to the lignocellulosic material, which is then processed at elevated temperatures and pressures, followed by rapid decompression. Generally, the temperatures are within 60–100 °C and the residence time does not exceed ten minutes. The major disadvantage is the large amount of ammonia used; approximately 1 kg of ammonia per kg of dry biomass. In addition, this process does not hydrolyze the biomass, but modifies its structure, allowing greater surface area for subsequent acid or enzymatic hydrolysis, resulting in lignocellulosic materials with high-lignin contents [95].

This process results in the depolymerization of cellulose and partial solubilization of hemicellulose and lignin. Thus, there is greater availability of fibers for hydrolysis. The sugars obtained from this pretreatment are well preserved with little or no degradation. Furthermore, using a closed reactor system, about 97% of the ammonia used in the process can be recovered and reused in subsequent pretreatments [96].

2.2.3.4 Supercritical CO_2 Explosion

A supercritical fluid has liquid and gas properties. This allows the fluid to penetrate the small pores of the lignocellulosic biomass. Typical pretreatment conditions are above the CO_2 critical temperature and pressure (31 °C and 1071 psi). Figure 2.4 illustrates the pretreatment experimental apparatus [98].

The biomass is usually moistened before pretreatment for better yields. The added water has two functions: formation of carbonic acid and to increase the biomass volume. The formation of carbonic acid favors the hydrolysis of the hemicellulose fraction. On the other hand, swelling opens the biomass pores to allow penetration of CO_2 and exposes a larger surface area of the material for enzymatic hydrolysis during rapid depressurization [97].

Supercritical CO_2 explosion seems to be a viable alternative due to its low-energy consumption, being also nontoxic, nonflammable, and inexpensive. Furthermore, this method does not generate harmful chemical residues and the

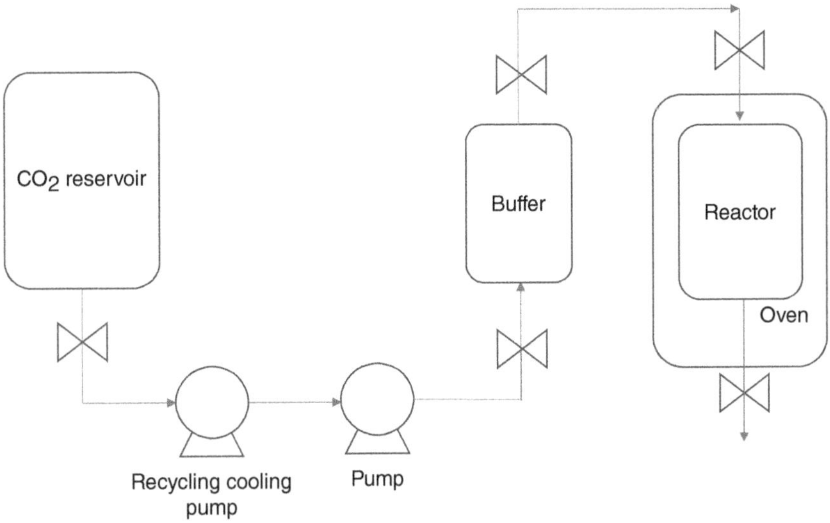

Figure 2.4 Supercritical CO_2 pretreatment. Source: Chundawat et al. [97]/with permission of Elsevier.

solvent can be easily removed from the products upon depressurization. However, it requires high-investment costs to construct the experimental setup needed for the supercritical CO_2 pretreatment. This is the major obstacle to its application on an industrial scale [99].

2.2.4 Biological Pretreatment

There is no inhibitory formation in the biological pretreatment process, which is less-energy intensive, being ecologically correct. Although fungi are microorganisms capable of effectively degrading lignin, they are not stable when using high temperatures and high pH values. On the other hand, several bacteria have been identified as capable of producing enzymes that degrade lignin and are used in the pretreatment of the lignocellulosic biomass. Among them are *Clostridium* sp., *Bacillus* sp., *Thermomonospora* sp., *Streptomyces* sp. *Trichoderma viride, Pseudomonas* sp., *Streptomyces* sp., *Aeromonas* sp., *Aspergillus niger*. Table 2.3 shows some literature work reporting biological pretreatments [116, 117].

The hydrolytic enzymatic system is responsible for the degradation of cellulose and hemicellulose, whereas the ligninolytic enzymatic system is capable of depolymerizing lignin through oxidative cleavage. Biological pretreatment technologies use microorganisms to solubilize lignin and remove it from the biomass material, without the need for adding chemicals or using high temperatures. The removal of lignin makes the cellulose structure more accessible, favoring the hydrolysis step [118, 119].

Table 2.3 Selected literature works on biological pretreatment applied for different biomasses.

Biomass	Microorganism	References
Sugarcane bagasse	*Ceriporiopsis submervispora*	[100]
Wheat straw	*Ceriporiopsis subvermispora*	[101]
Rice straw	*Pholiota adiposa*	[102]
Straw	*Fungal consortium*	[103]
Rice straw	*Bacillus firmus* K-1	[104]
Corn straw	*Mycobacterium smegmatis LZ-K2*	[105]
Corn stover	*Phanerochete chrysosporium*; *Coridus versicolor*	[106]
Bamboo culms	*Punctualaria sp. TUFC20056*	[107]
Eucalyptus *grandis*	*P. ostreatus; P. pulmonarius*	[108]
Corn stalks	*Irpex lacteus*	[109]
Corn stover	*Cyathus stercoreus* NRRL-6573	[110]
Corn stover	*Pycnoporus sanguineus* FP-10356-Sp	[110]
Corn stover	*Phlebia brevispora* NRRL-13108	[110]
Rice straw	MVI.2011	[111]
Rice hull	*Pleurotus ostreatus*	[112]
Rice straw	*Dichomitus squalens*	[113]
Rice straw	*Pleurotusostreatus*	[114]
Rubber wood	*C.subvermispora*	[115]

Despite being less energy-demanding, this process requires prolonged times and high-investment costs. The extremely low-hydrolysis rate is the main obstacle, which impact its industrial use; other factors include the need for careful growing conditions, large room space to carry out the biological pretreatment, and the required residence time of 10–14 days or even months [120–122].

2.3 Production of Levulinic Acid from Lignocellulosic Biomass

Different types of biomasses will present different yields for the same process. Thus, the choice of the biomass feedstock must consider its composition. Low concentrations of LA result in high separation costs; thus, it is important to choose the correct substrate. Furthermore, another important factor is the local availability, to

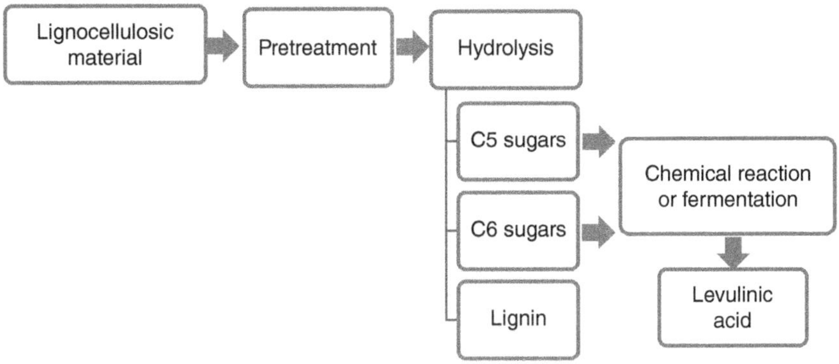

Figure 2.5 Main steps involved in the production of LA from lignocellulosic raw materials.

facilitate the creation of an efficient supply chain and logistics. Figure 2.5 shows the main steps involved in the production of LA from lignocellulosic feedstocks [123, 124].

The basic unit of the carbohydrates, named monosaccharides, such as glucose and fructose, can act as raw materials for LA synthesis. Fructose is easily dehydrated in 5-HMF using an acid catalyst – fructose is the best starting material for the 5-HMF synthesis, considering laboratory scale, whereas glucose needs to be first isomerized to fructose before conversion to 5-HMF. The presence of aldehyde and alcohol functional groups along the furan ring provides different patterns of reactivity, leading to the formation of variable compounds, as shown in Scheme 2.2. Once produced, 5-HMF can be converted into LA [125].

Stoichiometrically, 1 mol of LA can be produced from 1 mol of hexose. The rehydration of 5-HMF with two molecules of water leads to LA and formic acid. C6 sugars represent the ideal starting material for LA synthesis due to their high solubility in water and purity. Generally, two possible reaction conditions are used: (i) treatment with dilute acid at high temperature and atmospheric pressure; and (ii) treatment with highly concentrated acid at lower temperatures and atmospheric pressure [125, 126].

Furfural is another important precursor to sustainable LA production. Furfural may be obtained from pentoses, such as xylose and arabinose, through acid-catalyzed dehydration. In the synthesis process, furfural is catalytically reduced to furfuryl alcohol, followed by acid hydrolysis to produce LA, as shown in Scheme 2.3. Although concentrated acids are used as catalyst, the LA decomposition rate is faster than its production rate when the concentration of H_2SO_4 is greater than 10% by weight [10, 127].

In summary, due to the complexity of the lignocellulosic matrix, more than one technology may be needed. After the breakdown of the lignocellulosic biomass,

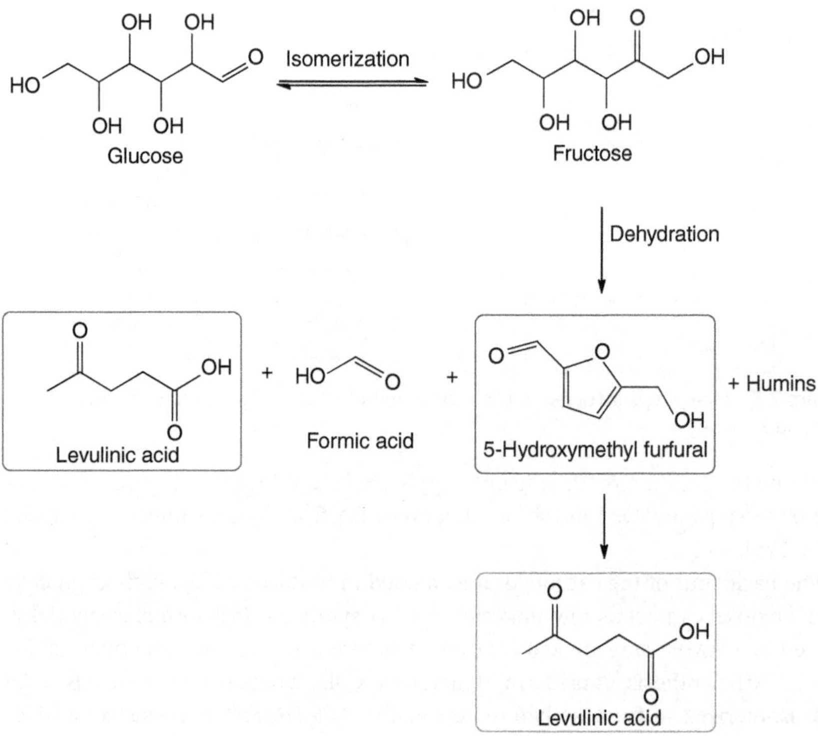

Scheme 2.2 Synthesis of 5-hydroxymethyl furfural and levulinic acid.

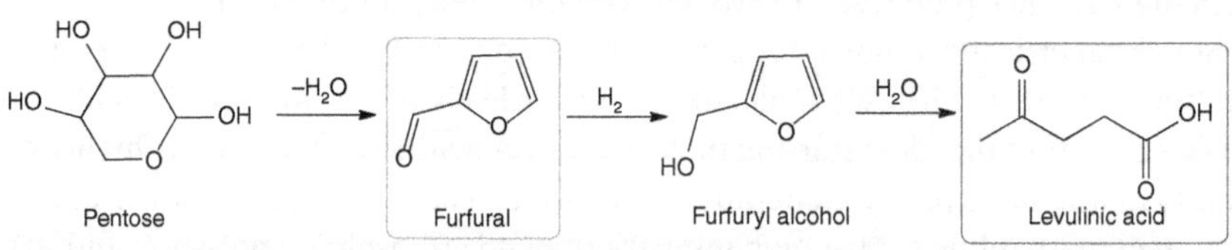

Scheme 2.3 Route of levulinic acid production from pentoses.

sugars are available for the conversion step, which can be carried out through chemical or biotechnological routes. However, research focused on the production of LA by biotechnological routes is still incipient. During the dehydration of sugars, furfural is the main product obtained from pentoses, whereas 5-HMF is generated upon the dehydration of hexose. Scheme 2.4 summarizes these routes [128].

The process of obtaining LA from biomass is carried out in aqueous phase using acid catalysts to release sugars and separate the lignin. At temperatures greater than 150 °C and under acidic conditions, the decomposition of carbohydrates can result in a variety of soluble products and humins, which represent the main solid

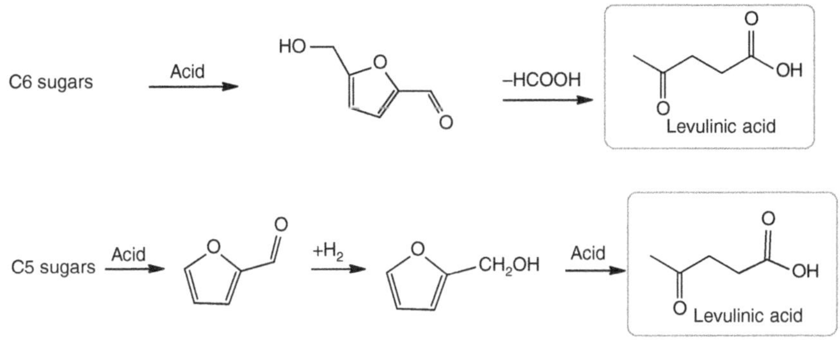

Scheme 2.4 Routes for the formation of LA from simple sugars.

products of the reaction. The formation of humins can lead to the loss of sugar, reducing the efficiency and economic viability of the process [129].

In summary, the hydrolysis of lignocellulosic biomass is complex, involving many reactions and intermediates (Scheme 2.5). In the presence of an acid catalyst and at elevated temperatures, the cellulose fraction is depolymerized to glucose, which is further converted to 5-HMF and finally to LA and formic acid. During hydrolysis, hemicellulose is depolymerized to pentoses and hexoses. The hexoses, such as glucose, galactose, and mannose, are converted to LA and formic acid as end products. Pentoses are converted to furfural, which is known to be quite reactive to form formic acid and other unwanted breakdown products. The lignin fraction in lignocellulosic biomass can be partially solubilized [74, 130–132].

As previously reported, the presence of lignin has a negative effect on the reaction yield, as lignin inhibits the conversion of cellulose into sugars. The production of formic acid and the formation of humins also negatively interfere with the reaction yield, causing blockage of the pipes and other equipment [131, 132].

Starch is considered one of the cheapest and most-abundant carbohydrates. However, it must be initially depolymerized upon hydrolysis for the formed glucose to be converted into 5-HMF, which requires longer reaction times or great amounts of acid catalysts to obtain yields similar to those found with monosaccharides. Finally, fructose has great potential, but the high cost of obtaining fructose prevents its application on an industrial scale [131, 132].

2.3.1 Processes for LA Production: Homogeneous Catalysts

The most-used homogeneous catalysts are minerals acids, such as H_2SO_4, HCl, and H_3PO_4, due to their low cost, good availability, and efficiency to achieve high-LA yields, although organic acids (acetic, oxalic, and citric acids), as well as Lewis acids ($AlCl_3$, $SnCl_4$, BF_3, SO_2, etc.) can also be used, affording good yields [74, 130–133].

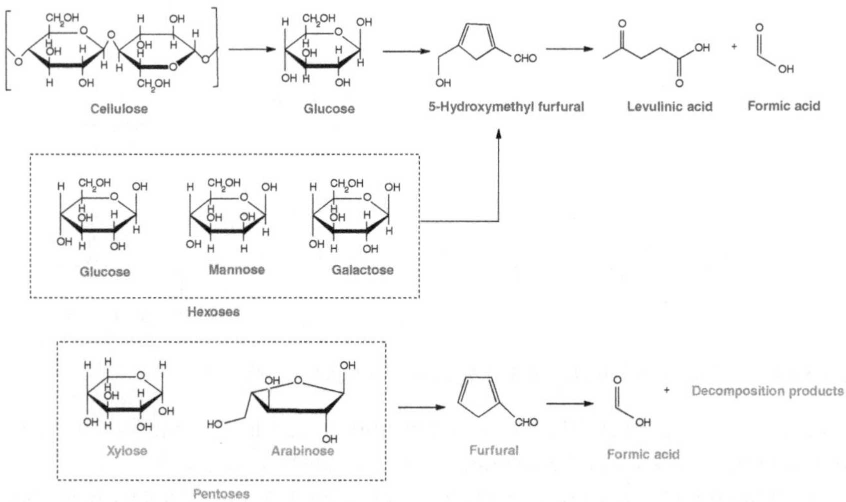

Scheme 2.5 Hydrolysis of sugars to LA and formic acid.

It is important to emphasize that mineral acids are difficult to be reused, leading to increased operational costs and generation of wastes. In addition, they may cause corrosion in the equipment, implying the use of special materials for the construction of the reactors. Thus, the capital and operational expenditures are increased. Table 2.4 shows some examples of LA production using homogeneous acid catalysts [131, 132].

Sulfuric acid was used in the hydrolysis of eucalyptus wood chips in the presence of methanol as solvent, at 180 °C, for 90 minutes. Under these conditions, 66 mol% of LA was produced, and the solvent inhibited the production of humins. The concentration of H_2SO_4 is a critical factor that affects product distribution [159, 160].

Brønsted acid catalysts, such as HCl, are required to catalyze dehydration and rehydration steps, whereas Lewis acids are needed for the isomerization of glucose to fructose. In this sense, some authors have focused their efforts on the synthesis of catalysts with good dispersion of acid sites, in search of greater efficiency in the process [161].

The separation and purification of LA from the acid aqueous solution is a great challenge. Complex purification steps are required to eliminate by-products. Extraction with organic solvents is the most commonly used method to separate LA from the medium. The solvent is then evaporated and can be recycled back to the extraction unit. Final purification of LA by vacuum distillation is also possible to improve the purity of the product [131–133, 159–161].

Table 2.4 Production of LA from biomass-derived feedstocks and homogeneous acid catalysts.

Substrate	Catalyst	Temperature (°C)	Time	Yield LA (wt%)	References
Glucose	HCl	160	4 h	41	[134]
Cellulose	H_2SO_4	150	2 h	43	[135]
Glucose	HCl	220	1 h	37.2	[136]
Frutose	H_3PO_4	240	0.03 h	4.5	[137]
Glucose	H_2SO_4	180	0.25 h	42	[138]
Glucose	H_2SO_4	98	12	24.5	[139]
Cellulose	HCl	180	20 min	44	[140]
Cellulose	$CrCl_3$	200	3 h	47.3	[141]
Glucose	HCl, $CrCl_3$	140	6	30	[142]
Wheat straw	H_2SO_4	209.3	37.6 min	19.86	[143]
Sorghum grain	H_2SO_4	200	40 min	32.6	[144]
Wood	HCl	250	2 h	12.4	[145]
Wood	HBr	200	4 h	8.1	[146]
Wood	H_2SO_4	200	4 h	15.5	[147]
Glucose	HCl	141	60 min	60	[140]
Glucose	H_2SO_4	180	15 min	42	[148]
Glucose	H_2SO_4	100	24 h	30	[149]
Fructose	HCl	140	80 min	72	[150]
Sugarcane bagasse	HCl	220	45 min	22.8	[151]
Olive tree pruning	HCl	200	1 h	47.2	[152]
Corn stover	H_2SO_4	160	19 h	66	[153]
Glucose	H_2SO_4	140–180	30–120 min.	30–60	[154]
Glucose	MSA	181.3	44.4 min	48.95	[155]
Eucalyptus wood	H_2SO_4	170	5 h	25.8	[156]
S. obliquus	HCl	180	10 min	45.63	[157]
Glucose	H_2SO_4	150	200 min	68.9	[158]
Brewery liquid waste	HCl	140	60 min	204*	[158]
Brewery spent grain	HCl	140	60 min	160*	[158]
Apple pomace solid wastes	HCl	140	60 min	66*	[158]
Apple pomace sludge	HCl	140	60 min	49*	[158]
Starch industry waste	HCl	140	60 min	12*	[158]

*(g Kg^{-1})

2.3.2 Processes for LA Production: Heterogeneous Catalysts

Solid acid catalysts are ideal to be used in industrial processes, as they can be easily separated from the reaction mixture for reuse. They can also be employed at high temperatures, and their acidity can be adjusted to improve the selectivity. Furthermore, heterogeneous catalysts are generally selective and do not present corrosion problems. Solid acid catalysts normally present Brønsted and Lewis acidity [162, 163].

Among the most common solid acids used in industrial processes are ion exchange resins, metal oxides, and zeolites. These materials have great advantages, because they can be reused, do not cause corrosion of the industrial equipment, and generate significantly fewer amounts of wastes [164]. Nevertheless, heterogeneous acid catalysts present some limitations when compared to homogeneous systems. First, the strong adsorption of reactants and products on the catalyst surface may decrease the overall LA yield. Many heterogeneous catalysts show deactivation upon reuse, which can be associated with two main factors: deposition or clogging of by-products, such as humins and lignin-derived residues, on the surface of the catalyst and leaching of the active phase. Although the yields obtained are lower when compared to homogeneous catalysis, proper process optimization can result in higher LA yields [163, 164].

The hydrolysis rate is dependent on the strength of the acid catalyst. Table 2.5 show some selected results. Although the LA yields can be promising in some cases, the values are not satisfactory for use on an industrial scale. The rates are usually slow and prolonged reaction times are necessary to obtain reasonable yields of LA. Therefore, developments are still needed to obtain an efficient heterogeneous catalyst that could replace the traditional minerals acids, especially for large-scale LA production [190].

The selectivity of zeolite catalysts may be associated with reactions that occur inside the pores and on the external surface. Zeolites ZSM-5, Beta, and Y were used to transform glucose into LA. However, the microporous structure limits the application of zeolite in converting larger substrates. In addition, zeolites can be used to replace resin catalysts because they can be regenerated upon calcination. Although LA production from furfuryl alcohol has been reported to give good yields over ion exchange resins, the regeneration of this type of catalyst upon deposition of solid humins is problematic [191, 192].

The most common method of catalyst regeneration is air calcination, in temperatures ranging from 400 to 500 °C. In cases where the material does not withstand high temperatures, the catalyst can be washed with solutions containing H_2O_2, HCl, NaOH, and others. Although some advances have been made in recent years regarding the use of heterogeneous catalysts for converting biomass to LA, the yields are still not competitive when compared with the yields obtained with mineral acids [131].

Table 2.5 Production of LA from biomass-derived feedstocks and heterogeneous catalysts.

Substrate	Catalysts	Temperature (°C)	Time	Yield LA (wt %)	References
Glucose	Al-Zr oxide	180	2 h	3.9	[165]
Glucose	Amberlyst 70	160	3 h	21.7	[134]
Cellulose	ZrO_2	180	3 h	39	[166]
Cellulose	Al-$NbOPO_4$	180	24 h	38	[167]
Cellulose	Amberlyst 70	160	16 h	69	[168]
Cellulose	ZrP	220	2 h	12	[134]
Cellulose	CP-SH	170	10 h	24	[169]
Corn stover	Amberlyst 70	160	16 h	54	[170]
Fructose	TFA; Ru/C	180	8 h	34	[171]
Glucose	CH_3-SBA-15-SO_3H	180	2.5 h	61.56	[172]
Fructose	sGO	160	1 h	61.2*	[173]
Glucose	$H_3O_{40}PW_{12}$	195	80 min	98.9	[174]
Glucose	Lys-PM_2	150	9 h	57.9	[175]
Glucose	Cr/NbP	180	180 min	62.4*	[176]
Agar	$KHSO_4$	115	6 h	56.48	[177]
Glucose	Cr/HZSM-5	180	180 min	64.4*	[178]
D-fructose	Dowex 50x8-100	120	24 h	72*	[179]
Cellulose	MCC20	180	10 min**	6	[180]
Corn straw	Amberlyst 15	180	30 min	~45***	[181]
Cellulose	Amberlyst 15	180	30 min	58.5***	[181]
Waste fluff	Amberlyst 15	180	30 min	~40***	[181]
Sorghum	Amberlyst 15	180	30 min	94.3***	[181]
Glucose	$CrCl_3$/HY zeolite	145.2	146.7 min	55.2	[182]
Chlorella sp.	$Al_2(SO_4)_3$	205	60 min	32.89	[183]
Corn stover	SAPO-18	190	80 min	70.20	[184]
Glucose	SBA-15-MPTMS	180	7 h	30	[185]
Fructose	SBA-15-MPTMS	180	7 h	16	[185]
Glucose	Fe/HY zeolite	180	3 h	62	[186]
Glucose	Ga-mordenite zeolite	175	6 h	59.9	[187]
Fructose	Amberlyst-15	120	24 h	52	[188]
D-fructose	LZY zeolite	140	15 h	43.2	[189]

* Mol%; ** Microwave heating; *** mg LA g^{-1} biomass.

The solvent is also important in heterogeneous acid catalysis, but the economic feasibility, easiness of separation and reuse, and environmental issues must be considered. Solvents, such as dimethyl sulfoxide (DMSO), dimethyl formamide (DMF), dimethyl acetamide (DMA), are not used on an industrial scale, because of the high costs, high-boiling points, and toxicity. The separation for reuse is carried out by distillation, especially for low boiling point solvents, such as tetrahydrofuran (THF). The best solvent for the sustainable synthesis of LA from sugars is water because it is greener, cheaper, and nontoxic [141, 193–195].

2.3.3 Processes for LA Production: Biphasic Systems

LA can also be produced in biphasic systems, involving water and an organic solvent. Both homogeneous and heterogeneous catalysts can be used and the product can be extracted to the organic phase. The use of an appropriate solvent can increase the solubility of cellulose, which increases the rate of mass transfer between the biomass feedstock and the catalyst, and consequently increases the conversion rate. Many studies report the use of two-phase reaction systems containing water and an aprotic polar solvent, such as THF, DMSO, and γ-valerolactone (GVL) [196, 197].

Water can solubilize carbohydrates and favors the rehydration of 5-HMF into levulinic and formic acids. Notwithstanding, it may also favor polymerization and formation of humins. Therefore, most works have focused on the search for catalysts that can increase the reaction yield and reduce side reactions [198].

Carbohydrates present low solubility in most common organic solvents. Volatile solvents are preferred because they are easier to be separated. Although DMSO is widely used, it is not recommended because of its high boiling point and toxicity [199].

The two-phase system allows the one-pot reaction and separation of the products. The aqueous phase solubilizes the carbohydrate and the catalyst, being the phase where the reaction takes place (Scheme 2.6). The formed product is extracted into the organic phase, in which its affinity is generally greater than with water. In addition, this system reduces cross-polymerization and rehydration, which mostly occur in the aqueous phase [74, 197, 200].

Generally, the hydrolysis of cellulose to glucose and its subsequent transformation to 5-HMF mainly occur in the aqueous phase. On the other hand, the conversion of 5-HMF in LA mostly occurs in the organic phase. To increase the partition coefficient between the two phases, it is common to use saturated salt solutions. For example, the solubilization of NaCl in water usually results in higher selectivity [197, 200].

The two-phase system composed of water and dimethoxymethane proved to be adequate for the synchronous conversion of hemicellulose and cellulose into LA,

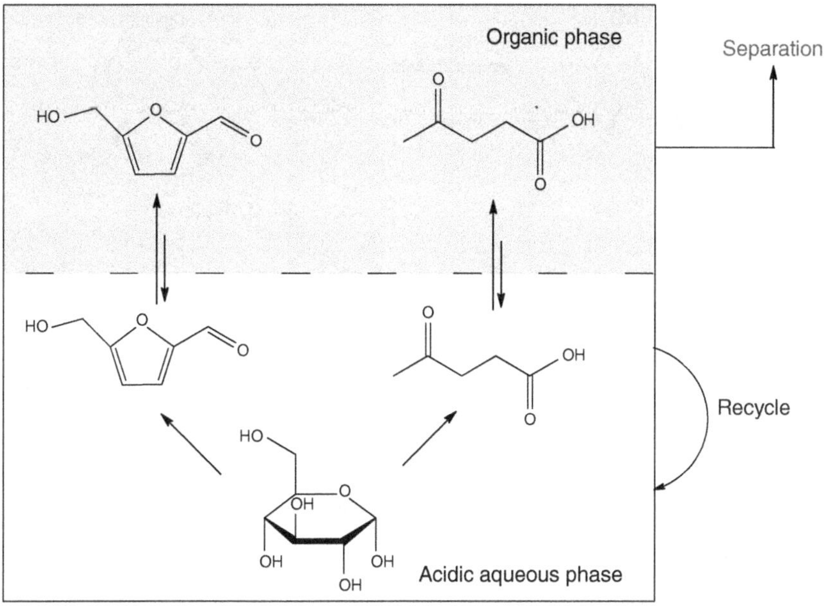

Scheme 2.6 Biphasic systems in LA production.

showing yields of 52.5% at 180 °C and 120 minutes. LA was produced in 15% yields from rice straw, using two-phase system composed of aqueous hydrochloric acid and dichloromethane at 180 °C for three hours. The solvent was recovered and recycled in up to five runs without significant loss of final product concentration [201, 202].

2.3.4 The Biofine Process of LA Production

Many commercial plants of LA production from biomass are based on the Biofine technology. The process consists of two acid-catalyzed steps, in which hexoses are converted to 5-HMF using mineral acid catalysts (1–4%) at temperatures ranging from 200 to 220 °C and pressures around 20 and 25 bar, for a few seconds. Then, the produced 5-HMF is transferred to a second reactor, where it is further hydrolyzed to LA at 190–220 °C and pressure of 10–15 bar for 15–30 minutes. Formic acid is a coproduct of the process, and the typical LA yield is around 70–80% (Scheme 2.7) [203, 204].

The process developed by Biofine Technology LLC, USA, comprises two differential reactors in sequence. The first converts the biomass into 5-HMF, whereas the second transforms the formed 5-HMF into LA. The reactor configuration minimizes the formation of coproducts and facilitates separation [203, 204].

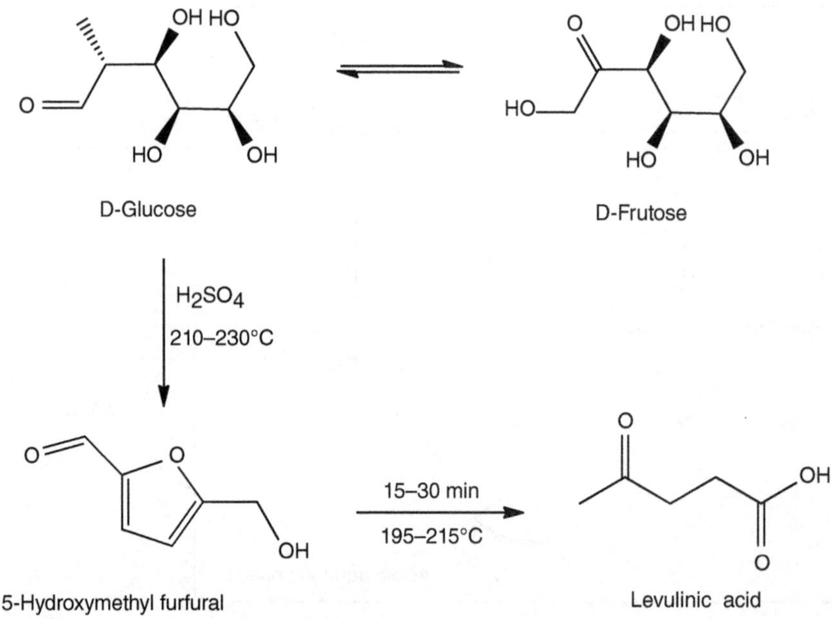

Scheme 2.7 Reactions of the Biofine process.

The main disadvantage of this process is the separation and recovery of the LA from the dilute aqueous solution. In addition, clogging of the piping and reactor systems can occur due to the formation of humins. A second point to be considered is the extensive neutralization and washing of the wastes, for their subsequent use in energy generation. A third point is the complex recovery of the mineral acid catalyst [203, 204].

A schematic representation of the process is shown in Figure 2.6. First, the carbohydrate feedstock and the sulfuric acid solution are mixed and fed into the reactor, operated at 210–220 °C, where they remain for an average residence time of 12 seconds to hydrolyze the polysaccharides into monomers and soluble sugar oligomers. The output of the first reactor is fed into a continuously stirred tank reactor (CSTR) operated at 190–200 °C, where the average residence time is 20 minutes. LA is separate in the liquid phase, whereas formic acid and furfural are recovered from the vapor stream phases. The solid by-products are separated from the aqueous solution of LA by filtration [132].

2.3.5 Downstream Process of LA Recovery

LA production can be carried out in batch or continuous flow. However, higher yields are obtained in continuous flow reactors. During the LA production process,

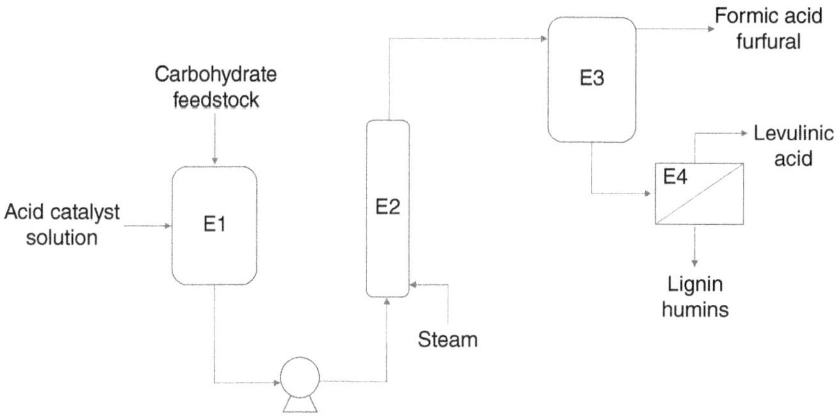

Figure 2.6 Schematic diagram of the Biofine process of levulinic acid production from sugars. E1: premixer; E2: tubular reactor; E3: tank reactor; E4: filter.

precipitation of mineral and organic salts can interfere in the conversion and separation due to clogging. Furthermore, the release of basic components can neutralize part of the acid catalyst [9].

The recovery processes, also called downstream, constitute a major challenge in the production of LA, due to formation of humins and the difficulty in separating LA from the inorganic acid catalyst. In the continuous process, humins are removed by filtration, followed by atmospheric or vacuum distillation and steam pickling, which can produce high-purity LA [205].

Different technologies can be applied in the LA separation and purification steps. The most-used technique in the continuous process is vacuum distillation, although it is energy-intensive. Solvent extraction and membrane separation can also be employed. Table 2.6 summarizes the methods employed with the respective advantages and disadvantages [205].

Solvent extraction is an expensive process at an industrial level due to the need of large volumes and the costs incurred in their recovery. The literature reports the use of furfural to extract LA and formic acid from the biomass acid hydrolysis. The main advantage is that furfural is relatively cheap and renewable, because it is obtained in the hydrolysis of pentoses. On the other hand, furfural presents toxicity, which makes the process not entirely green [206, 207].

The use of membranes to remove LA from the aqueous solutions is an interesting alternative due to its high selectivity and performance. In this process, the separation occurs because of the distinction in the diffusivity and solubility between the formed phases. The recovery of LA from fermentation broths or hydrolysates can also be done by adsorption through anion exchange resins or basic polymeric adsorbents [208–210].

Table 2.6 Advantages and limitations of different recovery processes for levulinic acid.

Recovery process	Advantages	Limitations
Vacuum distillation	Simple Well-established technology	Extensive energy requirements Formation of by-products
Solvent extraction	No additional processing step required	Large volumes of solvent required; toxicity problems Extraction of colored compounds
Steam stripping	High-purity product	Extensive energy requirements
Membrane separation	Single-step separation Minimizes undesired by-products	Expensive process Membrane fouling
Adsorption	Simple process	Low uptake at the industrial level
Ionic liquids	Still in research development	Expensive process Difficulty purification

Source: [205–212].

In an integrated production of furfural and LA, separation can take place by steam pickling. This is an energy-intensive process, and the residues are left behind, which may pollute the pickling tower. Due to the complexity of the chemical reactions or separation systems, combined separation methods are needed to improve the recovery or purification efficiency. For instance, ionic liquids have been used as solvents and catalysts. Upon completion of the reaction, ionic liquids can be separated from the products by extraction with organic solvents followed by distillation and filtration to separate the solid chemicals [211, 212].

2.4 Commercial Plants for the Production of LA

There are still few players operating the LA production on a commercial scale (Table 2.7). The North American company Biofine Technology is highlighted for having the largest share in the market and for developing the most-used technology. One of the advantages lies in the wide variety of lignocellulosic raw materials that can be used and, consequently, in the various LA derivatives that can be obtained. In 1987, Biofine Inc was founded, after receiving financial incentives from the US Department of Energy. In 2005, it was renamed Biofine Technologies LLC, an oil and energy services company. In 2017, the company built a new pilot plant in partnership with the University of Maine in Old Town. The process is

Table 2.7 Leading producers of levulinic acid in the world.

Company	Location	Production route
Biofine Technology LLC	United States	Biofine technology
GF Biochemicals Ltd.	Italy	Atlas technology
Langfang Hawk Technology & Development	China	ND
Bion-on	Italy	ND
Heroy Chemical Industry Co. Ltd.	China	ND
Parchem Fine & Specialty Chemicals	United States	ND
Avantium Inc.	Holland	ND
Hefei TNJ Chemical Industry Co. Ltd.	China	ND
Haihang Industry Co. Ltd.	China	ND
Tokyo Chemical Industry Co. Ltd.	Japan	ND

ND: Not available.
Source: [213, 214].

flexible in terms of the type of biomass used; any material with sufficient cellulose content and without an excess of ash may be considered [7, 213, 214].

According to information on the company's website, GF Biochemicals is the only company to produce LA derivatives on a commercial scale directly from biomass. GF Biochemicals expects substantial growth in the LA market in the forthcoming years. The technology has been developed for over 10 years and the company deposited over 200 patents.

GF Biochemicals was founded in 2008 to develop and commercialize LA through technology innovations. GF Biochemicals has developed a breakthrough technology to convert lignocellulosic biomass into LA. After the downstream acquisition of Segetis in 2016, the focus shifted to bringing the LA derivatives to the market. GF Biochemicals has a large portfolio of products, going from LA to its main derivatives, such as the levulinate esters. The technology is flexible in terms of feedstock and enables cost-competitive production.

The first commercial-scale plant for the synthesis of LA was built in Caserta, Italy, using the technology developed by Biofine Renewables (Figure 2.7). GF Biochemicals is considered the main producer of LA from renewable sources, with a projected annual production capacity of 50 000 tons between the years 2020 and 2025. LA production may use cornstarch as raw material. In 2016, the process was changed to use cellulosic biomass, such as wood residues, grass, and wheat straw, through acid pretreatment and heating [215, 216].

The production process consists of a pretreatment step using acid hydrolysis to separate the C5 and C6 sugars. The reactor is charged with inorganic acid and

Figure 2.7 GF Biochemicals LA plant in Italy. Source: GFBiochemicals Ltd.

heated to 160 °C. LA is produced at high temperatures and pressures. The system is cooled down and the solid and liquid fractions are separated. The separation process occurs with organic solvent extraction [217–219].

Langfang Hawk Technology & Development is a Chinese company founded in 1992 consisting of two research and development centers, four subsidiary plants, and a multifunctional pilot plant. The company produces LA through its subsidiary founded in 2002. The production capacity is 3 Kt/year, using furfural as raw material and inorganic acid catalysis. In addition to LA, the company also sells some derivatives, such as α-angelicalactone, γ-valerolactone, methyl levulinate, ethyl levulinate, propyl levulinate, butyl levulinate, calcium levulinate, sodium levulinate, diphenolic acid, azopolymerization initiators, dimethyl sulfite and others products (http://www.hawk-chinachem.com/web/index.asp).

In 2017, Bion-on started a project to produce LA on a commercial scale, at a competitive price, and with low-environmental impact. The project aims to use by-products of the sugar industry as raw material, with an annual production capacity of 5 Kt. Bion-on stands out in the production of biopolymers, actively participating in the dissemination of new technologies for the production of polyhydroxyalkanoates (PHAs), a biological plastic produced from agricultural processing residues, such as sugar beet and sugarcane (http://www.bio-on.it/index.php).

Heroy Chemical Industry Co. is a Chinese company that produces LA since 1995. Today, in addition to LA, the company produces methyl levulinate, ethyl levulinate, propyl levulinate, butyl levulinate, among other derivatives (http://www.heroychem.com/index.html).

Parchem is a company located in New Rochelle, New York, specialized in the production of items from various segments, such as food, beverage, sweeteners, coatings, adhesives sealants, cosmetics, personal care, flavor, fragrance, pharmaceuticals, in addition to the production of LA and its derivatives, calcium and sodium levulinate, among others (https://www.parchem.com/index.aspx).

Avantium is a Dutch company that invests in the production of 100% vegetable, recyclable, and degradable plastics. The company produces methyl levulinate, methoxymethyl-furfural (MMF), furandicarboxylic acid (FDCA), and polyethylene furanoate (PEF). Chinese Hefei TNJ Chemical Industry Co. Ltd. produces LA and other intermediate products as well as pharmaceuticals, solvents, and vitamins, among others (https://www.avantium.com/, https://pt.tnjchem.com/about-us_d10).

Haihang Industry Co. Ltd. is a Chinese manufacturer of flavorings, fragrances, food additives, pharmaceuticals, and others. It sells some levulinate ester derivatives, such as methyl levulinate, ethyl levulinate, butyl levulinate, and calcium levulinate. The Japanese company Tokyo Chemical Industry Co.Ltd. trades LA, methyl levulinate, ethyl levulinate, propyl levulinate, butyl levulinate, 5-aminolevulinic acid hydrochloride, methyl 5-aminolevulinate hydrochloride, and α and β-angelicalactone (https://haihangchem.com/?gclid=CjwKCAiA5t-OBhByEiwAhR-hm9CnOFF2LtVsTl0xYEcTU3IYPn3Fm2y-Krc7i-w454LKYKmzI-R7bxoCrwEQAvD_BwE, https://www.tcichemicals.com/JP/en/).

2.5 Conclusion

Although great advances have been made in the production of LA from biomass, some issues still need to be improved. The use of heterogeneous catalysts is not yet feasible on an industrial scale because of the formation of solid humin by-products and leaching of active sites. In this sense, the development of cheaper, selective, thermally stable, and easily recyclable heterogeneous catalysts is still a major challenge.

The use of raw and residual biomass is necessary for the sustainable production on an industrial scale. In addition, the recovery and use of solid wastes is another important aspect to be addressed from a sustainable biorefinery perspective.

References

1 Farnleitner, L., Stueckler, H., Kaiser, H., and Kloimstein, E. (1991). Preparation of stable levulinic acid. German patent 3920340 to Chemie Linz G.M.B.H.

2 Itaya, H., Shiotani, A., Toriyahara, Y. (1988). Preparation of levulinic acid from furfuryl alcoholo. Japanese patent 62252742 to Ube Industries, Ltd., issued 4 November 1987.
3 Edwards, W.B. (1986). III Oxocarboxylic acids. US patent 4612391 to Philip Morris, Inc., issued 16 September 1986.
4 Vaerman, J. M.; Bertrand, J. (1972). Selective oxidation of ketones. German patent 2125162 to Labofina SA.
5 Cavinato, G. and Toniolo, L. (1990). Levulinic acid synthesis via regiospecific carbonylation of methyl vinyl ketone or of its reaction products with hydrochloric acid or an alkanol or of a mixture of acetone with a formaldehyde precursor catalyzed by a highly active palladium-hydrochloric acide. *J. Mol. Catal.* 58: 251–267.
6 Ballini, R. and Petrini, M. (1986). Facile and inexpensive synthesis of 4-oxoalkanoic acids from primary nitroalkanes and acroleine. *Synthesis* 1024–1026.
7 De Souza, R.O.M.A., Miranda, L.S.M., and Luque, R. (2014). Bio(chemo)technological strategies for biomass conversion into bioethanol and key carboxylic acids. *Green Chem.* 16: 2386–2405.
8 Morales-Delarosa, S. and Campos-Martin, J.M. (2014). *Catalytic Processes and Catalyst Development in Biorefining*. Spain: Woodhead Publishing Limited, Instituto de Catálisis y Petroleoquímica, CSIC.
9 Wettstein, S.G., Alonso, D.M., and Gürbüz, E.I. (2012). A roadmap for conversion of lignocellulosic biomass to chemicals and fuels. *Curr. Opin. Chem. Eng.* 1: 218–224.
10 Adeleye, A.T., Louis, H., Akakuru, O.U. et al. (2019). A review on the conversion of levulinic acid and its esters to various useful chemicals. *AIMS Energy* 2: 165–185.
11 Negus, M.P., Mansfield, A.C., and Leadbeater, N.E. (2015). The preparation of ethyl levulinate facilitated by flow processing: the catalyzed and uncatalyzed esterification of levulinic acid. *J. Flow Chem.* 3: 148–150.
12 Kang, S., Fu, J., and Zhang, G. (2018). From lignocellulosic biomass to levulinic acid: a review on acid-catalyzed hydrolysis. *Renewable Sustainable Energy Rev.* 94: 340–362.
13 Antonetti, C., Licursi, D., Fulignati, S. et al. (2016). New frontiers in the catalytic synthesis of levulinic acid: from sugars to raw and waste biomass as starting feedstock. *Catalyysis* 6 (196): 1–29.
14 Sanderson, K. (2011). Lignocellulose: a chewy problem. *Nature* 474: 12–14.
15 Rodrigues, C., Woiciechowski, A.L., Letti, L.A.J. et al. (2017). Materiais lignocelulósicos como matéria-prima para a obtenção de biomoléculas de valor comercial. *Biotecnologia Aplicada à Agro&Indústria* 8: 283–314.

16 Baruah, J., Nath, B.K., Sharma, R. et al. (2018). Recent trends in the pretreatment of lignocellulosic biomass for value-added products. *Front. Energy Res.* 6: 1–19.

17 Usmani, Z., Sharma, M., Gupta, P. et al. (2020). Ionic liquid based pretreatment of lignocellulosic biomass for enhanced bioconversion. *Bioresour. Technol.* 304: 123003.

18 Cardona, C.A., Quintero, J.A., and Paz, I.C. (2010). Production of bioethanol from the sugar cane bagasse: status and perspectives. *Bioresour. Technol.* 101: 4754–4760.

19 Jeffries, T.W. (1990). Biodegradation of lignin-carbohydrate complexes. *Biodegradation* 1: 163–176.

20 Li, X., Xu, R., Yang, J. et al. (2019). Production of 5-hydroxymethylfurfural and levulinic acid from lignocellulosic biomass and catalytic upgradation. *Ind. Crop, Prod.* 130: 184–197.

21 Kumar, S., Singh, S.P., Mishra, I.M., and Adhikari, D.K. (2009). Recent advances in production of bioethanol from lignocellulosic biomass. *Chem. Eng. Technol* 32: 517–526.

22 Behera, S., Arora, R., Nandhagopal, N., and Kumar, S. (2014). Importance of chemical pretreatment for bioconversion of lignocellulosic biomass. *Renew. Sust. Energ. Rev.* 36: 91–106.

23 Li, J., Yang, Z., Zhang, K. et al. (2021). Valorizing waste liquor from dilute acid pretreatment of lignocellulosic biomass by *Bacillus megaterium* B-10. *Ind. Crop, Prod.* 161: 113160.

24 Mosier, N., Wyman, C., Dale, B. et al. (2005). Features of promising technologies for pretreatment of lignocellulosic biomass. *Bioresour. Technol.* 96: 673–686.

25 Agbor, B.V., Cicek, N., Sparling, R. et al. (2011). Biomass pretreatment: fundamentals toward application. *Biotechnol. Adv.* 29: 675–685.

26 Ravindran, R. and Jaiswal, A.K. (2016). A comprehensive review on pre-treatment strategy for lignocellulosic food industry waste: challenges and opportunities. *Bioresour. Technol.* 199: 92–10293.

27 Mood, S.H., Golfeshan, A.H., Tabatabaei, M. et al. (2013). Lignocellulosic biomass to bioethanol, a comprehensive review with a focus on pretreatment. *Renewable Sustainable Energy Rev.* 27: 77–93.

28 Hendricks, A.T. and Zeeman, G. (2009). Pretreatments to enhance the digestibility of lignocellulosic biomass. *Bioresour. Technol.* 100: 10–18.

29 Zakaria, M.R., Norrrahim, M.N.F., Hirata, S., and Hassan, M.A. (2015). Hydrothermal and wet disk milling pretreatment for high conversion of biosugars from oil palm mesocarp fiber. *Bioresour. Technol.* 181: 263–269.

30 Gu, B.J., Wang, J., Wolcott, M.P., and Ganjyal, G.M. (2018). Increased sugar yield from pre-milled Douglas-fir forest residuals with lower energy consumption by using planetary ball milling. *Bioresour. Technol.* 251: 93–98.

31 Kim, H.J., Lee, S., Kim, J. et al. (2013). Environmentally friendly pretreatment of plant biomass by planetary and attrition milling. *Bioresour. Technol.* 144: 50–56.

32 Negro, M.J., Duque, A., Manzanares, P. et al. (2015). Alkaline twin-screw extrusion fractionation of olive-tree pruning biomass. *Ind. Crops Prod.* 74: 336–341.

33 Wang, Z., Hou, X., Sun, J. et al. (2018). Comparison of ultrasound-assisted ionic liquid and alkaline pretreatment of Eucalyptus for enhancing enzymatic saccharification. *Bioresour. Technol.* 254: 145–150.

34 Li, H., Qu, Y., Yang, Y. et al. (2016). Microwave irradiation–A green and efficient way to pretreat biomass. *Bioresour. Technol.* 199: 34–41.

35 Fatriasari, W., Fajriutami, T., Laksana, R.B., and andWistara, N. J. (2018). Microwave assisted-acid hydrolysis of jabon kraft pulp. *Waste Biomass Valorization* 9: 1–15.

36 Verma, A., Kumar, S., and Jain, P.K. (2011). Key pretreatment technologies on cellulosic ethanol production. *J. Sci. Res.* 55: 57–63.

37 Ooshima, H., Aso, K., Harano, Y., and Yamamoto, T. (1984). Microwave treatment of cellulosic materials for their enzymatic hydrolysis. *Biotechnol. Lett.* 6: 289.

38 Liyakathali, N.A.M., Muley, P.D., Aita, G., and Boldor, D. (2016). Effect of frequency and reaction time in focused ultrasonic pretreatment of energy cane bagasse for bioethanol production. *Bioresour. Technol.* 200: 262–271.

39 Luo, J., Fang, Z., and Smith, R.L. Jr., (2014). Ultrasound-enhanced conversion of biomass to biofuels. *Prog. Energy Combust. Sci.* 41: 56–93.

40 Nitayavardhana, S.K., Rakshit, D.G., Leeuwen, J.V., and Khanal, S.K. (2008). Ultrasound pretreatment of cassava chip slurry to enhance sugar release for sub-sequent ethanol production. *Biotechnol. Bioeng.* 101: 487–496.

41 Nguyen, Q.A., Tucker, M.P., Keller, F.A., and Eddy, F.P. (2000). Two-stage dilute-acid pretreatment of softwoods. *Appl. Biochem. Biotechnol.* 84–86: 561–575.

42 Jönsson, L.J. and Martín, C. (2016). Pretreatment of lignocellulose: formation of inhibitory by-products and strategies for minimizing their effects. *Bioresour. Technol.* 199: 103–112.

43 Liu, C., Lu, X., Yu, Z. et al. (2020). Production of levulinic acid from cellulose and cellulosic biomass in different catalytic systems. *Catalysis* 10: 1–22.

44 Sivers, M.V. and Zacchi, G.A. (1995). Techno-economical comparison of three processes for the production of ethanol from pine. *Bioresour Technol.* 51: 43–52.

45 Hamelinck, C.N., Hooijdonk, G.V., and Faaij, A.P.C. (2005). Ethanol from lignocellulosic biomass: techno-economic performance in short-, middle- and long-term. *Biomass Bioenergy* 28: 384–410.

46 Chandra, R.P., Bura, R., Mabee, W.E. et al. (2007). Substrate pretreatment: the key to effective enzymatic hydrolysis of lignocellulosics? *Adv. Biochem. Eng. Biotechnol.* 108: 67–93.

47 Maurya, D.P., Vats, S., Rai, S., and Negi, S. (2013). Optimization of enzymatic saccharification of microwave pretreated sugarcane tops through response surface methodology for biofuel. *Indian J. Exp. Biol.* 51: 992–996.

48 Sakuragi, K., Igarashi, K., and Samejima, M. (2018). Application of ammonia pretreatment to enable enzymatic hydrolysis of hardwood biomass. *Polym. Degrad. Stab.* 148: 19–25.

49 Chang, V.S., Kaar, W.E., Burr, B., and Holtzapple, M.T. (2001). Simultaneous saccharifcation and fermentation of lime-treated biomass. *Biotech Lett.* 16: 1327–1233.

50 Shetty, D.J., Kshirsagar, P., Tapadia-Maheshwari, S. et al. (2017). Alkali pretreatment at ambient temperature: a promising method to enhance biomethanation of rice straw. *Bioresour. Technol.* 226: 80–88.

51 Talha, Z., Ding, W., Mehryar, E. et al. (2016). Alkaline pretreatment of sugarcane bagasse and filter mud codigested to improve biomethane production. *Biomed. Res. Int.* 1–10.

52 Alayoubi, R., Mehmood, N., Husson, E. et al. (2020). Low temperature ionic liquid pretreatment of lignocellulosic biomass to enhance bioethanol yield. *Renewable Energy* 145: 1808–1816.

53 Costa, S.P.F., Azevedo, A.M.O., Pinto, P.C.A.G., and Saraiva, M.L. (2017). Environmental impact of ionic liquids: recent advances in (Eco)toxicology and (Bio) degradability. *ChemSusChem* 11: 2321–2347.

54 Semerci, I. and Güler, F. (2018). Protic ionic liquids as effective agents for pretreatment of cotton stalks at high biomass loading. *Ind. Crops Prod.* 125: 588–595.

55 Hu, X., Cheng, L., Gu, Z. et al. (2018). Effects of ionic liquid/water mixture pretreatment on the composition, the structure and the enzymatic hydrolysis of corn stalk. *Ind. Crops Prod.* 122: 142–147.

56 Li, Q., He, Y.C., Xian, M. et al. (2009). Improving enzymatic hydrolysis of wheat straw using ionic liquid 1-ethyl-3-methyl imidazolium diethyl phosphate pretreatment. *Bioresour. Technol.* 14: 3570–3575.

57 Rocha, E.G.A., Pin, T.C., Rabelo, S.C., and Costa, A.C. (2017). Evaluation of the use of protic ionic liquids on biomass fractionation. *Fuel* 206: 145–154.

58 Liu, Z., Li, L., Liu, C., and Xu, A. (2018). Pretreatment of corn straw using the alkaline solution of ionic liquids. *Bioresour. Technol.* 260: 417–420.

59 Hu, L., Peng, H., Xia, Q. et al. (2020). Effect of ionic liquid pretreatment on the physicochemical properties of hemicellulose from bamboo. *J. Mol. Struct.* 1210: 1–8.

60 Pin, T.C., Nakasu, P.Y.S., Mattedi, S. et al. (2019). Screening of protic ionic liquids for sugarcane bagasse pretreatment. *Fuel* 235: 1506–1514.

61 Mohammadi, M., Shafiei, M., Abdolmaleki, A. et al. (2019). A morpholinium ionic liquid for rice straw pretreatment to enhance ethanol production. *Ind. Crop. Prod.* 139: 1–9.

62 Zhang, K., Wei, L., Sun, Q. et al. (2021). effects of formaldehyde on fermentable sugars production in the low-cost pretreatment of corn stalk based on ionic liquids. *Chin. J. Chem. Eng.* 23.

63 Hou, X., Wang, Z., Sun, J. et al. (2019). A microwave-assisted aqueous ionic liquid pretreatment to enhance enzymatic hydrolysis of *Eucalyptus* and its mechanism. *Bioresour. Technol.* 272: 99–104.

64 Tan, H.T. and Lee, K.T. (2012). Understanding the impact of ionic liquid pretreatment on biomass and enzymatic hydrolysis. *Chem. Eng. J.* 183: 448–458.

65 Papa, G., Feldman, T., Sale, K.L. et al. (2017). Parametric study for the optimization of ionic liquid pretreatment of corn stover. *Bioresour. Technol.* 241: 627–637.

66 Aung, E.M., Endo, T., Fujii, S. et al. (2018). Efficient pretreatment of bagasse at high loading in an ionic liquid. *Ind. Crops Prod.* 11: 243–248.

67 Fu, D. and Mazza, G. (2011). Aqueous ionic liquid pretreatment of straw. *Bioresour. Technol.* 102: 7008–7011.

68 Fu, D. and Mazza, G. (2011). Optimization of processing conditions for the pretreatment of wheat straw using aqueous ionic liquid. *Bioresour. Technol.* 102: 8003–8010.

69 Farahani, S.V., Kim, Y.W., and Schall, C.A. (2016). A coupled low temperature oxidative and ionic liquid pretreatment of lignocellulosic biomass. *Catal. Today* 269: 2–8.

70 Reis, C.L.B., Silva, L.M.A., Rodrigues, T.H.S. et al. (2017). Pretreatment of cashew apple bagasse using protic ionic liquids: enhanced enzymatic hydrolysis. *Bioresour. Technol.* 224: 694–701.

71 Holm, J., Lassi, U., Romar, H. et al. (2012). Pretreatment of fibre sludge in ionic liquids followed by enzyme and acid catalysed hydrolysis. *Catal. Today* 196: 11–15.

72 Perez-Pimienta, J.A., Lopez-Ortega, M.G., Chavez-Carvayar, J.A. et al. (2015). Characterization of agave bagasse as a function of ionic liquid pretreatment. *Biomass Bioenergy* 75: 180–188.

73 Yadav, N., Nain, L., and Khare, S.K. (2021). One-pot production of lactic acid from rice straw pretreated with ionic liquid. *Bioresour. Technol.* 323: 124563.

74 Rackemann, D.W. and Doherty, W.O.S. (2011). The conversion of lignocellulosics to levulinic acid. *Biofuels, Bioprod. Biorefin* 5: 198–214.

75 Yoo, C.G., Pu, Y., and Ragauskas, A.J. (2017). Ionic liquids: promising green solvents for lignocellulosic biomass utilization. *Curr. Opin. Green Sustainable Chem.* 5: 5–11.

76 Auxenfans, T., Buchoux, S., Larcher, D. et al. (2014). Enzymatic saccharification and structural properties of industrial wood sawdust: recycled ionic liquids pretreatments. *Energy Convers. Manage.* 88: 1094–1103.

77 Nguyen, T.A.D., Kim, K.R., Han, S.J. et al. (2010). Pretreatment of rice straw with ammonia and ionic liquid for lignocellulose conversion to fermentable sugars. *Bioresour. Technol.* 101: 7432–7438.

78 Teramura, H., Sasaki, K., Oshima, T. et al. (2018). Effective usage of sorghum bagasse: optimization of organosolv pretreatment using 25% 1-butanol and subsequent nanofiltration membrane separation. *Bioresour. Technol.* 252: 157–164.

79 Meng, X., Bhagia, S., Wang, Y. et al. (2020). Effects of the advanced organosolv pretreatment strategies on structural properties of woody biomass. *Ind. Crops Prod.* 146: 1121442.

80 Zhang, K., Pei, Z., and Wang, D. (2016). Organic solvent pretreatment of lignocellulosic biomass for biofuels and biochemicals: a review. *Bioresour. Technol.* 199: 21–33.

81 Baral, N.R. and Shah, A. (2017). Comparative techno-economic analysis of steam explosion, dilutesulfuric acid, ammonia fiber explosion and biological pretreatments of corn stover. *Bioresour. Technol.* 232: 331–343.

82 Stoffel, R.B., Neves, P.V., Felissia, F.E. et al. (2017). Hemicellulose extraction from slash pine sawdust by steam explosion with sulfuric acid. *Biomass Bioenergy* 107: 93–10194.

83 Lorenzo-Hernando, A., Martín-Juárez, J., and Bolado-Rodríguez, S. (2018). Study of steam explosion pretreatment and preservation methods of commercial celulose. *Carbohydr. Polym.* 191: 234–241.

84 Jacquet, N., Quiévy, N., Vanderghem, C. et al. (2011). Influence of steam explosion on the thermal stability of cellulose fibres. *Polym. Degrad. Stab.* 9: 1582–1588.

85 Silva, T.A.L., Zamora, H.D., Varão, L.H.R. et al. (2018). Effect of steam explosion pretreatment catalysed by organic acid and alkali on chemical and structural properties and enzymatic hydrolysis of sugarcane bagasse. *Waste Biomass* 9: 2191–2201.

86 Baêta, B.E.L., Cordeiro, P.H.M., Passos, F. et al. (2017). Steam explosion pretreatment improved the biomethanization of coffeehusks. *Bioresour. Technol.* 245: 7266–7268.

87 Yu, Y., Jie, J., Ren, X. et al. (2022). Steam explosion of lignocellulosic biomass for multiple advanced bioenergy processes: a review. *Renewable Sustainable Energy Rev.* 154: 111871.

88 Martín, C., Wu, G., Wang, Z. et al. (2018). Formation of microbial inhibitors in steam-explosion pretreatment of softwood impregnated with sulfuric acid and sulfur dioxide. *Bioresour. Technol.* 262: 242–250.

89 Ao, T., Luo, Y., Chen, Y. et al. (2020). Towards zero waste: a valorization route of washing separation and liquid hot water consecutive pretreatment to achieve solid vinasse based biorefinery. *J. Cleaner Prod.* 248: 119253.

90 Saksit Imman, S., Arnthong, J., Burapatana, V. et al. (2014). Effects of acid and alkali promoters on compressed liquid hot water pretreatment of rice straw. *Bioresour. Technol.* 171: 29–36.

91 Tian, D., Shen, F., Yang, G. et al. (2019). Liquid hot water extraction followed by mechanical extrusion as a chemical-free pretreatment approach for cellulosic ethanol production from rigid hardwood. *Fuel* 252: 589–597.

92 Jimenez-Gutierrez, J.M., Van Der Wielen, L.A.M., and Straathof, A.J.J. (2020). Subcritical CO_2 shows no effect on liquid hot water pretreatment of poplar wood. *Bioresour. Technol.* 11: 100442.

93 Imman, S., Kreetachat, T., Khongchamnan, P. et al. (2021). Optimization of sugar recovery from pineapple leaves by acid-catalyzed liquid hot water pretreatment for bioethanol production. *Energy Rep.* 7: 6945–6954.

94 Tian, S., Zhang, H., and Fu, S. (2022). Improvement of xylo-oligosaccharides dissolution from *Caragana korshinskii* through liquid hot water pretreatment with tiny choline chloride. *Ind. Crops Prod.* 176: 114418.

95 Chundawat, S.P.S., Chang, L., Gunawan, C. et al. (2012). Guayule as a feedstock for lignocellulosic biorefineries using ammonia fiber expansion (AFEX) pretreatment. *Ind. Crops Prod.* 37: 486–492.

96 Abdul, P.M., Jahim, J.M., Harun, S. et al. (2016). Effects of changes in chemical and structural characteristic of ammonia fibre expansion (AFEX) pretreated oil palm empty fruit bunch fibre on enzymatic saccharification and fermentability for biohydrogen. *Bioresour. Technol.* 211: 200–208.

97 Zha, M.J., Xu, Q.Q., Li, G.M. et al. (2019). Pretreatment of agricultural residues by supercritical CO_2 at 50–80 °C to enhance enzymatic hydrolysis. *J. Energy Chem.* 31: 39–45.

98 Gu, T. (2013). *Pretreatment of Lignocellulosic Biomass Using Supercritical Carbon Dioxide as a Green Solvent. Green Biomass Pretreatment for Biofuels Production, Chapter 5*, 107–125. Springer.

99 Escobar, E.L.N., da Silva, T.A., Pirich, C.L. et al. (2020). Supercritical fluids: a promising technique for biomass pretreatment and fractionation. *Front. Bioeng. Biotechnol.* 8: 1–18.

100 Machado, A.D.S. and Ferraz, A. (2017). Biological pretreatment of sugarcane bagasse with basidiomycetes producing varied patterns of biodegradation. *Bioresour. Technol.* 225: 17–22.

101 Nayan, N., Sonnenberg, A.S.M., Hendriks, W.H., and Cone, J.W. (2018). Screening of white-rot fungi for bioprocessing of wheat straw into ruminant feed. *J. Appl. Microbiol.* 125: 468–479.

102 Jagtap, S.S., Dhiman, S.S., Kim, T.S. et al. (2013). Characterization of a β-1, 4-glucosidase from a newly isolated strain of *Pholiota adiposa* and its application to the hydrolysis of biomass. *Biomass Bioenergy* 54: 181–190.

103 Taha, M., Shahsavari, E., Hothaly, K.A. et al. (2015). Enhanced biological straw saccharification through coculturing of lignocellulose-degrading microorganisms. *Appl. Biochem. Biotechnol.* 175: 3709–3728.

104 Baramee, S., Siriatcharanon, A.K., Ketbot, P. et al. (2020). Biological pretreatment of rice straw with cellulase-free xylanolytic enzyme-producing *Bacillus firmus* K-1: Structural modification and biomass digestibility. *Renewable Energy* 160: 555–563.

105 Zhang, K., Xu, R., Abomohra, A.E.F. et al. (2019). A sustainable approach for efficient conversion of lignin into biodiesel accompanied by biological pretreatment of corn straw. *Energy Convers. Manage.* 199: 111928.

106 Wang, F.Q., Xie, H., Chen, W. et al. (2013). Biological pretreatment of corn stover with ligninolytic enzyme for high efficient enzymatic hydrolysis. *Bioresour. Technol.* 144: 572–578.

107 Suhara, H., Kodama, S., Kamei, I. et al. (2012). Screening of selective lignin-degrading basidiomycetes and biological pretreatment for enzymatic hydrolysis of bamboo culms. *Int. Biodeter. Biodegr.* 75: 176–180.

108 Castoldi, R., Bracht, A., Morais, G.R. et al. (2014). Biological pretreatment of *Eucalyptus grandis* sawdust with white-rot fungi: study of degradation patterns and saccharification kinetics. *Chem. Eng. J.* 258: 240–246.

109 Du, W., Yu, H., Song, L. et al. (2011). The promising effects of by-products from *Irpex lacteus* on subsequent enzymatic hydrolysis of bio-pretreated corn stalks. *Biotechnol. Biofuels* 4: 37.

110 Saha, B.C., Qureshi, N., Kennedy, G.J., and Cotta, M.A. (2016). Biological pretreatment of corn stover with white-rot fungus for improved enzymatic hydrolysis. *Int Biodeter. Biodegr.* 109: 29–35.

111 Sreemahadevan, S., Roychoudhury, P.K., Thankamani, V., and Ahammad, S.Z. (2018). Biological pretreatment of rice straw using an alkalophilic fungus MVI.2011 for enhanced enzymatic hydrolysis yield. *Sustainable Energy Technol. Assess.* 30: 304–313.

112 Yu, J., Zhang, J., He, J. et al. (2009). Combinations of mild physical or chemical pretreatment with biological pretreatment for enzymatic hydrolysis of rice hull. *Bioresour. Technol.* 100: 903–908.

113 Bak, J.S., Kim, M.D., Choi, I.G., and Kim, K.H. (2010). Biological pretreatment of rice straw by fermenting with *Dichomitus squalens*. *New Biotechnol.* 27: 424–434.

114 Taniguchi, M., Suzuki, H., Watanabe, D. et al. (2005). Evaluation of pretreatment with *Pleurotus ostreatus* for enzymatic hydrolysis of rice straw. *J. Biosci. Bioeng.* 100: 637–643.

115 Nazarpour, F., Abdullah, D.K., Abdullah, N., and Zamiri, R. (2013). Evaluation of biological pretreatment of rubberwood with white rot fungi for enzymatic hydrolysis. *Materials* 6: 2059–2073.

116 Vats, S., Maurya, D.P., Shaimoon, M. et al. (2013). Development of a microbial consortium for the production of blend enzymes for the hydrolysis of agricultural waste into sugars. *J. Sci. Ind. Res.* 72: 585–590.

117 Wan, C. and Li, Y. (2012). Fungal pretreatment of lignocellulosic biomass. *Biotechnol. Adv.* 6: 1447–1157.

118 Howard, R.L., Abotsi, E.L., and Howard, S. (2003). Lignocellulose biotechnology issues of bioconversion and enzyme production. *Afr. J. Biotechnol.* 12: 602–619.

119 Mtui, G.Y.S. (2009). Recent advances in pretreatment of lignocellulosic wastes and production of value added products. *Afr. J. Biotechnol.* 8: 1398–1415.

120 Magnusson, L., Islam, R., Sparling, R. et al. (2008). Direct hydrogen production from cellulosic waste materials with a single-step dark fermentation process. *Int. J. Hydrog. Energy* 33: 5398–5403.

121 Wan, C. and Li, Y. (2011). Effect of hot water extraction and liquid hot water pretreatment on the fungal degradation of biomass feedstocks. *Bioresour. Technol.* 102: 9788–9793.

122 Sodre, V., Vilela, N., Tramontina, R., and Squina, F.M. (2021). Microorganisms as bioabatement agents in biomass to bioproducts application. *Biomass Bioenergy* 151: 10616.

123 Galbe, M. and Wallberg, O. (2019). Pretreatment for biorefneries: a review of common methods for efficient utilisation of lignocellulosic materials. *Biotechnol. Biofuels* 12: 1–26.

124 Kumar, A., Shende, D.Z., and Wasewar, K.L. (2020). Production of levulinic acid: a promising building block material for pharmaceutical and food industry. *Mater. Today:. Proc.* 29: 790–793.

125 Morone, A., Apte, M., and Pandey, R.A. (2015). Levulinic acid production from renewable waste resources: bottlenecks, potential remedies, advancements and application. *Renewable Sustainable Energy Rev.* 51: 548–565.

126 Assary, R.S., Redfern, P.C., and Hammond, J.R. (2010). Computational studies of the thermochemistry for conversion of glucose to levulinic acid. *J. Phys. Chem. B* 114: 9002–9009.

127 Yan, L., Greenwood, A.A., Hossain, A., and Yang, B. (2014). The comprehensive mechanistic kinetic model for dilute acid hydrolysis of switchgrass cellulose to glucose, 5-HMF and levulinic acid. *RSC Adv.* 4: 23492–23504.

128 Girisuta, B., Dussan, K., Haverty, D. et al. (2013). A kinetic study of acid catalysed hydrolysis of sugar cane bagasse to levulinic acid. *Chem. Eng. J.* 217: 61–70.

129 Szabolcs, A., Molnár, M., Dibó, G., and Mika, L.T. (2013). Microwave-assisted conversion of carbohydrates to levulinic acid: an essential step in biomass conversion. *Green Chem.* 15: 439–445.

130 Elumalai, S., Agarwal, B., and Sangwan, R.S. (2016). Thermo-chemical pretreatment of rice straw for further processing for levulinic acid production. *Bioresour. Technol.* 21: 232–246.

131 Covinich, L.G., Clauser, N.M., Felissia, F.E. et al. (2019). The challenge of converting biomass polysacharides into levulinic acid through heterogeneous catalytic processes. *Biofuels, Bioprod. Biorefin.* 14: 1–29.

132 Girisuta, B. and Heeres, H.J. (2017). Chapter 5 - Levulinic acid from biomass: synthesis and applications. Production of platform chemicals from sustainable resources. *Biofuel, Biorefin.* 7: 143–169.

133 Singh, M., Pandey, N., Dwivedi, P. et al. (2019). Production of xylose, levulinic acid, and lignin from spent aromatic biomass with a recyclable Brønsted acid synthesized from d-limonene as renewable feedstock from citrus waste. *Bioresour. Technol.* 293: 1221053.

134 Weingarten, R., Conner, W.C., and Huber, G.W. (2012). Production of levulinic acid from cellulose by hydrothermal decomposition combined with aqueous phase dehydration with a solid acid catalyst. *Energy Environ. Sci.* 5: 7559–7574.

135 Girisuta, B., Janssen, L.P.B.M., and Heeres, H.J. (2007). Kinetic study on the acid-catalyzed hydrolysis of cellulose to levulinic acid. *Ind. Eng. Chem. Res.* 46: 1696–1708.

136 Deng, L., Li, J., Lai, D.-M. et al. (2009). Catalytic conversion of biomass-derived carbohydrates into-valerolactone without using an external H_2 Supply. *Angew. Chem. Int. Ed.* 48: 6529–6532.

137 Salak, A.F. and Yoshida, H. (2006). Acid-catalyzed production of 5-hydroxymethyl furfural from D-fructose in subcritical water. *Ind. Eng. Chem. Res.* 45: 2163–2173.

138 Rackemann, D.W., Bartley, J.P., and Doherty, W.O.S. (2014). Methanesulfonic acid-catalyzed conversion of glucose and xylose mixtures to levulinic acid and furfural. *Ind. Crops Prod.* 52: 46–57.

139 Tarabanko, V.E., Chernyak, M.Y., Aralova, S.V., and Kuznetsov, B.N. (2002). Kinetics of levulinic acid formation from carbohydrates at moderate temperatures. *React. Kinet. Catal. Lett.* 75: 117–126.

140 Shen, J. and Wyman, C.E. (2012). Hydrochloric acid-catalyzed levulinic acid formation from cellulose: data and kinetic model to maximize yields. *AIChE J.* 58: 236–246.

141 Peng, L., Lin, L., Zhang, J. et al. (2010). Catalytic conversion of cellulose to levulinic acid by metal chlorides. *Molecules* 15: 5258–5272.

142 Choudhary, V., Mushrif, S.H., Ho, C. et al. (2013). Insights into the interplay of lewis and brønsted acid catalysts in glucose and fructose conversion to 5-(hydroxymethyl)furfural and levulinic acid in aqueous media. *J. Am. Chem. Soc.* 135: 3997–4006.

143 Chang, C., Cen, P., and Ma, X. (2007). Levulinic acid production from wheat straw. *Bioresour. Technol.* 98: 1448–1453.

144 Fang, Q. and Hanna, M. (2002). Experimental studies for levulinic acid production from whole kernel grain sorghum. *Bioresour Technol.* 3: 187–192.

145 Efremov, A., Pervyshina, G., and Kuznetsov, B. (1997). Thermocatalytic transformations of wood and cellulose in the presence of HCl, HBr, and H_2SO_4. *Chem Nat Compd* 33: 84–88.

146 Girisuta B. (2007) Levulinic acid from lignocellulosic biomass. Chemical Engineering [Ph.D. thesis]. Netherlands: University of Groningen.

147 Rackemann, D.W., Bartley, J.P., and Doherty, W.O.S. (2014). Methanesulfonic acid-catalysed conversion of glucose and xylose mixtures to levulinic acid and furfural. *Ind. Crops Prod.* 52: 46–57.

148 Brasholz, M., von Känel, K., Hornung, C.H. et al. (2011). Highly efficient dehydration of carbohydrates to 5-(chloromethyl)furfural (CMF), 5- (hydroxymethyl)furfural (HMF) and levulinic acid by biphasic continuous flow processing. *Green Chem.* 13: 1114–1117.

149 Yan, L., Yang, N., Pang, H., and Liao, B. (2008). Production of levulinic acid from bagasse and paddy straw by liquefaction in the presence of hydrochloride acid. *CLEAN - Soil Air Water* 36: 158–163.

150 Galletti, A.M.R., Antonetti, C., Luies, V. et al. (2012). Levulinic acid from waste. *Bioresour.* 7: 1824–1835.

151 Alonso, D.M., Wettstein, S.G., Mellmer, M.A. et al. (2013). Integrated conversion of hemicellulose and cellulose from lignocellulosic biomass. *Energy Environ. Sci.* 1: 76–80.

152 Mikola, M., Ahola, J., and Tanskanen, J. (2019). Production of levulinic acid from glucose in sulfolane/water mixtures. *Chem. Eng. Res. Des.* 148: 291–297.

153 Kim, H.S., Kim, S.K., and Jeong, G.T. (2018). Catalytic conversion of glucose into levulinic and formic acids using aqueous Brønsted acid. *J. Ind. Eng. Chem.* 63: 48–56.

154 Kang, S. and Yu, J. (2016). An intensified reaction technology for high levulinic acid concentration from lignocellulosic biomass. *Biomass Bioenergy* 95: 214–220.

155 Jeong, G.T. and Kim, S.K. (2020). Valorization of thermochemical conversion of lipid-extracted microalgae to levulinic acid. *Bioresou. Technol.* 313: 123684.

156 Lau, K.S., Chin, S.X., Jaafar, S.N.S., and Chia, C.H. (2021). Conversion of glucose into levulinic acid in continuous segmented turbulent flow with enhanced chemical reaction. *Tetrahedron Lett.* 80: 15333.

157 Maiti, S., Gallastegui, G., Suresh, G. et al. (2018). Microwave-assisted one-pot conversion of agro-industrial wastes intolevulinic acid: an alternate approach. *Bioresour. Technol.* 265: 471–479.

158 Chen, S.S., Maneerung, T., Tsang, D.C.W. et al. (2017). Valorization of biomass to hydroxymethylfurfural, levulinic acid, and fatty acid methyl ester by heterogeneous catalysts. *Chem. Eng. J.* 328: 246–273.

159 Kang, S. and Yu, J. (2015). Effect of methanol on formation of levulinates from cellulosic biomass. *Ind. Eng. Chem. Res.* 54: 11552–11559.

160 Kang, S., Fu, J., Zhou, N. et al. (2018). Concentrated levulinic acid production from sugar cane molasses. *Energy Fuels* 32: 3526–3531.

161 Garcés, D., Faba, L., Díaz, E., and Ordóñez, S. (2019). Aqueous phase transformation of glucose into HMF and levulinic acid combining homogeneous and heterogeneous catalysis. *ChemSusChem* https://doi.org/10.1002/cssc.201802315.

162 Stöcker, M. (2008). Biofuels and biomass-to-liquid fuels in the biorefinery: catalytic conversion of lignocellulosic biomass using porous materials. *Angew. Chem. Int. Ed. Engl.* 47: 9200–9211.

163 Tang, P. and Yu, J. (2014). Kinetic analysis on deactivation of a solid Brønsted acid catalyst in conversion of sucrose to levulinic acid. *Ind. Eng. Chem. Res.* 53: 11629–11637.

164 Lima, E.E., Silva, A.S., Silva, F.L.H. et al. (2012). Hidrólise de celulose por catalisadores mesoestruturados NiO-MCM-41 E MoO_3-MCM-41. *Quim. Nova* 35: 683–688.

165 Zeng, W., Cheng, D.G., Chen, F., and Zhan, X. (2009). Catalytic conversion of glucose on Al–Zr mixed oxides in hot compressed water. *Catal. Lett.* 133: 221–226.

166 Joshi, S.S., Zodge, A.D., Pandare, K.V., and Kulkarni, B.D. (2014). Efficient conversion of cellulose to levulinic acid by hydrothermal treatment using zirconium dioxide as a recyclable solid acid catalyst. *Ind. Eng. Chem. Res.* 53: 18796–18805.

167 Ding, D., Wang, J., Xi, J. et al. (2014). High-yield production of levulinic acid from cellulose and its upgrading to -valerolactone. *Green Chem.* 16: 3846–3853.

168 Alonso, D.M., Gallo, J.M.R., Mellmer, M.A. et al. (2013). Direct conversion of celulose to levulinic acid and gamma-valerolactone using solid acid catalysts. *Catal. Sci. Technol.* 3: 927–931.

169 Zuo, Y., Zhang, Y., and Fu, Y. (2014). Catalytic conversion of cellulose into levulinic acid by a sulfonated cholormethyl polystyrene solid acid catalyst. *ChemCatChem* 6: 753–757.

170 Alonso, D.M., Gallo, J.M.R., Mellmer, M.A. et al. (2013). Direct conversion of cellulose to levulinic acid and gamma-valerolactone using solid acid catalysts. *Catal. Sci. Technol.* 3: 927–931.

171 Heeres, H., Handana, R., Chunai, D. et al. (2009). Combined dehydration/(transfer)-hydrogenation of C-6 sugars (D-glucose and D-fructose) to γ-valerolactone using ruthenium catalysts. *Green Chem.* 11: 1247–1255.

172 Cheng, X., Feng, Q., Ma, D. et al. (2021). Efficient catalytic production of levulinic acid over hydrothermally stable propyl sulfonic acid functionalized SBA-15 in γ-valerolactone-water system. *J. Environ. Chem. Eng.* 9: 105747.

173 Lawagon, C.P., Faungnawakij, K., Srinives, S. et al. (2021). Sulfonated graphene oxide from petrochemical waste oil for efficient conversion of fructose into levulinic acid. *Catal. Today* 375: 197–203.

174 Li, X., Lu, X., Nie, S. et al. (2020). Efficient catalytic production of biomass-derived levulinic acid over phosphotungstic acid in deep eutectic solvente. *Ind. Crops Prod.* 145: 112154.

175 Qu, H., Liu, B., Gao, G. et al. (2019). Metal-organic framework containing Brønsted acidity and Lewis acidity for efficient conversion glucose to levulinic acid. *Fuel Process. Technol.* 193: 1–6.

176 Wei, W., Yang, H., and Wu, S. (2019). Efficient conversion of carbohydrates into levulinic acid over chromium modified niobium phosphate catalyst. *Fuel* 256: 115940.

177 Kholiya, F., Rathod, M.R., Gangapura, D.R. et al. (2020). An integrated effluent free process for the production of 5-hydroxymethyl furfural (HMF), levulinic acid (LA) and KNS-ML from aqueous seaweed extract. *Carbohydr. Res.* 490: 107953.

178 Weiqi, W. and Shubin, W. (2018). Experimental and kinetic study of glucose conversion to levulinic acid in aqueous medium over Cr/HZSM-5 catalyst. *Fuel* 225: 311–321.

179 Thapa, I., Mullen, B., Saleem, A. et al. (2017). Efficient green catalysis for the conversion of fructose to levulinic acid. *Appl. Catal.* 539: 70–79.

180 Ching, T.W., Haritos, V., and Tanksale, A. (2017). Microwave assisted conversion of microcrystalline cellulose into value added chemicals using dilute acid catalyst. *Carbohydr. Polym.* 157: 1794–1800.

181 Nis, B. and Ozsel, B.K. (2020). A comprehensive experimental study on the production of key platform chemicals from raw biomass. *Fuel* 280: 118674.

182 Ya'aini, N., Amin, N.A.S., and Asmadi, M. (2012). Optimization of levulinic acid from lignocellulosic biomass using a new hybrid catalyst. *Bioresour. Technol.* 116: 58–65.

183 Jeong, G.T. and Kim, S.K. (2021). Hydrothermal conversion of microalgae *Chlorella* sp. into 5-hydroxymethylfurfural and levulinic acid by metal sulfate catalyst. *Biomass Bioenergy* 148: 106053.

184 Li, X., Xu, R., Liu, Q. et al. (2019). Valorization of corn stover into furfural and levulinic acid over SAPO-18 zeolites: effect of Brønsted to Lewis acid sites ratios. *Ind. Crops Prod.* 141: 111759.

185 Pizzolitto, C., Ghedini, E., Menegazzo, F. et al. (2020). Effect of grafting solvent in the optimisation of SBA-15 acidity for levulinic acid production. *Catal. Today* 345: 183–189.

186 Ramli, N.A.S. and Amin, N.A.S. (2015). Fe/HY zeolite as an effective catalyst for levulinic acid production from glucose: characterization and catalytic performance. *Appl. Catal. B Environ.* 163: 487–498.

187 Kumar, V.B., Pulidindi, I.N., Mishra, R.K., and Gedanken, A. (2016). Ga modified zeolite based solid acid catalyst for levulinic acid production. *ChemistrySelect* 1: 5952–5960.

188 Son, P.A., Nishimura, S., and Ebitani, K. (2012). Synthesis of levulinic acid from fructose using Amberlyst-15 as a solid acid catalyst. *React. Kinet. Mech. Catal.* 106: 185–192.

189 Jow, J., Rorrer, G.L., and Hawley, M.C. (1987). Dehydration of D fructose to levulinic acid over LZY zeolite catalyst. *Biomass* 14: 185–194.

190 Xiang, M., Liu, J., Fu, W. et al. (2017). Improved activity for cellulose conversion to levulinic acid through hierarchization of ETS-10 Zeolite. *ACS Sustainable Chem. Eng.* 5: 5800–5809.

191 Gürbüz, E.I., Wettstein, S.G., and Dumesic, J.A. (2021). Conversion of hemicellulose to furfural and levulinic acid using biphasic reactors with alkylphenol solvents. *ChemSusChem* 5: 383–387.

192 Pratama, A.P., Rahayu, D.U.C., and Krisnandi, Y.K. (2020). Levulinic acid production from delignified rice husk waste over manganese catalysts: heterogeneous versus homogeneous. *Catalysis* 10: 1–13.

193 Dutta, S., Yu, I.K.M., Tsang, D.C.W. et al. (2020). Influence of green solvent on levulinic acid production from lignocellulosic paper waste. *Bioresour. Technol.* 298: 122544.

194 Tsilomelekis, G., Josephson, T.R., Nikolakis, V., and Caratzoulas, S. (2014). Origin of 5-hydroxymethylfurfural stability in water/dimethyl sulfoxied mixtures. *ChemSusChem* 7: 117–126.

195 Morone, A., Apte, M., and Panday, R.A. (2015). Levulinic acid production from renewable waste resources: bottlenecks, potential remedies, advancements and applications. *Renewable and Sustainable Energy Rev.* 51: 548–565.

196 Wettstein, S.G., Alonso, D.M., Chonga, Y., and Dumesic, J.A. (2012). Production of levulinic acid and gamma-valerolactone (GVL) from cellulose using GVL as a solvent in biphasic systems. *Energy Environ. Sci.* 5: 8199–8203.

197 Ma, C., Cai, B., Zhang, L. et al. (2021). Acid-catalyzed conversion of cellulose into levulinic acid with biphasic solvent system. *Front. Plant Sci.* 12: 1–10.

198 Carvalho, E.G.L., Rodrigues, F.A., Monteiro, R.S. et al. (2018). *Experimental Design and Economic Analysis of 5-hydroxymethylfurfural Synthesis from Fructose in Acetone-water System Using Niobium Phosphate as Catalyst*, Biomass Conversion and Biorefinery. Springer.

199 Jeong, J., Antonyraj, C.A., Shin, S. et al. (2013). Commercially attractive process for production of 5-hydroxymethyl-2-furfural from high fructose corn syrup. *J. Ind. Eng. Chem.* 19: 1106–1111.

200 Dutta, S., Yu, I.K.M., Tsang, C.W. et al. (2020). Influence of green solvente on levulinic acid production from lignocellulosic paper waste. *Bioresour. Technol.* 298: 12254.

201 Feng, J., Tong, L., Xu, Y. et al. (2020). Synchronous conversion of lignocellulosic polysaccharides to levulinic acid with synergic bifunctional catalysts in a biphasic cosolvent system. *Ind. Crop. Prod.* 145: 112084.

202 Kumar, S., Ahluwalia, V., Kundu, P. et al. (2018). Improved levulinic acid production from agri-residue biomass in biphasicsolvent system through synergistic catalytic effect of acid and products. *Bioresour. Technol.* 251: 143–115.

203 Bozell, J.J., Moens, L., Elliott, D.C. et al. (2000). Production of levulinic acid and use as a platform chemical for derived products. *Resour. Conserv. Recycl.* 28: 227–239.

204 Kapanj, K.K., Haigh, K.F., and Gorgens, J.F. (2021). Techno-economics of lignocellulose biorefineries at South African sugar mills using the biofine process to co-produce levulinic acid, furfural and electricity along with gamma valeractone. *Biomass Bioenergy* 146: 106008.

205 Pileidis, F.D. and Titirici, M.M. (2016). Levulinic acid biorefineries: new challenges for efficient utilization of biomass. *ChemSusChem* 9: 562–582.

206 Nhien, L.C., Long, N.V.D., and Lee, M. (2016). Design and optimization of the levulinic acid recovery process from lignocellulosic biomass. *Chem. Eng. Res. Des.* 107: 126–136.

207 Kumar, A., Shende, D.Z., and Wasewar, K.L. (2020). Extractive separation of levulinic acid using natural and chemical solvents. *Chem. Data Collect.* 28: 100417.

208 Baylan, N. and Çehreli, S. (2018). Ionic liquids as bulk liquid membranes on levulinic acid removal: a design study. *J. Mol. Liq.* 266: 299–308.

209 Liu, B. and Liu, S. (2011). Adsorption of levulinic acid from aqueous solution onto basic polymeric adsorbent: experimental and modeling studies. *Sep. Sci. Technol.* 46: 2391–2399.

210 Datta, D. and Uslu, H. (2017). Adsorption of levulinic acid from aqueous solution by Amberlite XAD-4. *J. Mol. Liq.* 234: 330–334.

211 Ji, H., Zhu, J.Y., and Roland Gleisner, R. (2017). Integrated production of furfural and levulinic acid from corncob in a one-pot batch reaction incorporating distillation using step temperature profiling. *RSC Adv.* 7: 46208–46214.

212 Zhou, J., Sui, H., Jia, Z. et al. (2018). Recovery and purification of ionic liquids from solutions: a review. *RSC Adv.* 8: 32832–32864.

213 Gozan, M., Ryan, B., and Krisnandi, Y. (2018). Techno-economic assessment of levulinic acid plant from sorghum bicolor in Indonesia. *IOP Conf. Ser.: Mater. Sci. Eng.* 345 (012012).

214 Lima, M.T. (2017) Desenvolvimento de químicos intermediários de base renovável: os casos do ácido levulínico e itacônico. Rio de Janeiro.

215 Hayes, D.J., Fitzpatrick, S., Hayes, M.H.B., and Ross, J.R.H. (2008). The biofine process – production of levulinic acid, furfural, and formic acid from lignocellulosic feedstocks. In: *Biorefineries-Industrial Processes and Products*, 139–164. Wiley-VCH Verlag GmbH.

216 Ramli, N.A.S. and Amin, N.A.S. (2016). Kinetic study of glucose conversion to levulinic acid over Fe/HY zeolite catalyst. *Chem. Eng. J.* 283: 150–159.

217 Girisuta, B., Danon, B., Manurung, R. et al. (2008). Experimental and kinetic modelling studies on the acid-catalysed hydrolysis of the water hyacinth plant to levulinic acid. *Bioresour. Technol.* 99: 8367–8375.

218 Ritter, S. (2006). Biorefinery gets ready to deliver the goods. *Chem. Eng. News* 47.

219 Mullen, B.D., Yontz, D.J., and Leibig, C.M. (2017). Process to prepare levulinic acid. Patent US 9,598,341 B2. GFBiochemicals Limited, issued 21 March 2017.

3

Levulinate Derivatives – Main Production Routes and Uses of Organic and Inorganic Levulinates Derivatives

3.1 Main Production Routes

Alkyl levulinate (AL) esters and levulinate salts (LS) are chemicals with great potential for industrial applications. When obtained from renewable sources (Figure 3.1), they become interesting alternatives to replace current chemical products produced from petrochemical routes [1, 2].

The levulinate esters can be synthesized by two main routes: (i) esterification of levulinic acid, and (ii) direct production from the alcoholysis of polysaccharides. Both routes can be seen in Scheme 3.1.

3.1.1 Esterification of Levulinic Acid

The most traditional route of AL esters production consists of the classical esterification of levulinic acid with alcohols (Scheme 3.2).

Since esterification is reversible, high levulinic acid conversions can be achieved by shifting the chemical equilibrium through water removal during the reaction, or upon the use of molar excess of the alcohol. However, the reactivity of the carboxyl group is relatively low, and long reaction times, as well as high temperatures and pressures, are usually required to achieve the equilibrium conversion in the absence of catalysts [4, 5]. The use of homogeneous or heterogeneous acid catalysts accelerates the reaction, permitting to achieve high selectivity and yield of the AL ester [2, 4]. Homogeneous catalytic processes present numerous disadvantages. For example, strong mineral acids cause environmental damage and corrosion problems. Thus, the search for heterogeneous acid catalysts, especially those that can be easily separated, recycled, regenerated, or reused, is gaining importance [2]. A variety of solid acid catalysts have been studied for the esterification of levulinic acid. Table 3.1 shows some literature results.

Different types of materials have been evaluated in the esterification of levulinic acid to AL esters. Amberlyst-15 resin, for example, was tested and the yields of

Levulinic Acid: A Sustainable Platform Chemical for Value-Added Products, First Edition.
Claudio J.A. Mota, Ana Lúcia de Lima, Daniella R. Fernandes, and Bianca P. Pinto.

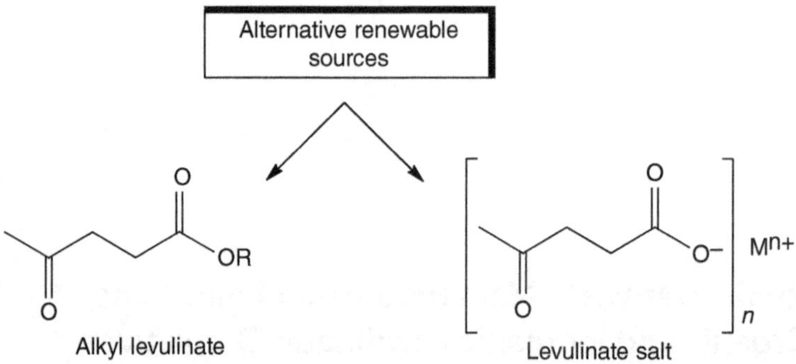

Figure 3.1 Chemical structure of levulinate derivatives, where R is an alkyl group and M^{n+} stands for a metal cation.

organic levulinates ranged from 54 to 84%, depending on the type of alcohol, molar ratio, catalyst loading, and temperature [6–8].

Fernandes et al. studied the conversion of levulinic acid to ethyl levulinate on some acid zeolite catalysts. The conversion depends on the structure of the zeolite [6]. Authors proposed two mechanistic pathways, depending on the dimension of the zeolite pores or cavities. The first pathway involves the protonation of the carbonyl oxygen atom of levulinic acid, which enables the further nucleophilic attack by the alcohol to form a bulkier tetrahedral intermediate, featuring a quaternary carbon atom. This pathway can explain the results on large-pores zeolites. The other pathway involves protonation of the hydroxyl group of levulinic acid, followed by water elimination to afford a linear acyl cation intermediate, which demand less space to be formed inside the pores. This mechanistic route may explain the results in more constrained zeolite systems. In both cases, protonation of levulinic acid and the formation of the respective intermediates can be assisted by the zeolite structure. The authors suggested that while the proton is being transferred to the oxygen atom of the carbonyl, the framework oxygen atom may nucleophilically assist the carbon atom to help delocalizing the positive charge (Figure 3.2). Nevertheless, this spatial arrangement may impose some steric constraints between the carbonyl substituent groups and the zeolite framework. The sp2 hybridization of the carbonyl moiety imposes a planar geometry with angles of 120°. Thus, upon coordination of the carbonyl oxygen atom with the acidic site, the substituent groups are pointed toward the opposite side and may interact with the framework structure. In addition, the carbon atom changes hybridization from sp2 to sp3, which demands more space to accommodate the substituents. Therefore, the steric repulsion will be significantly more or less severe depending on the pore constraint of the structure and protonation of the carbonyl oxygen atom may not be favored. By contrast, protonation on the hydroxyl group may form a highly

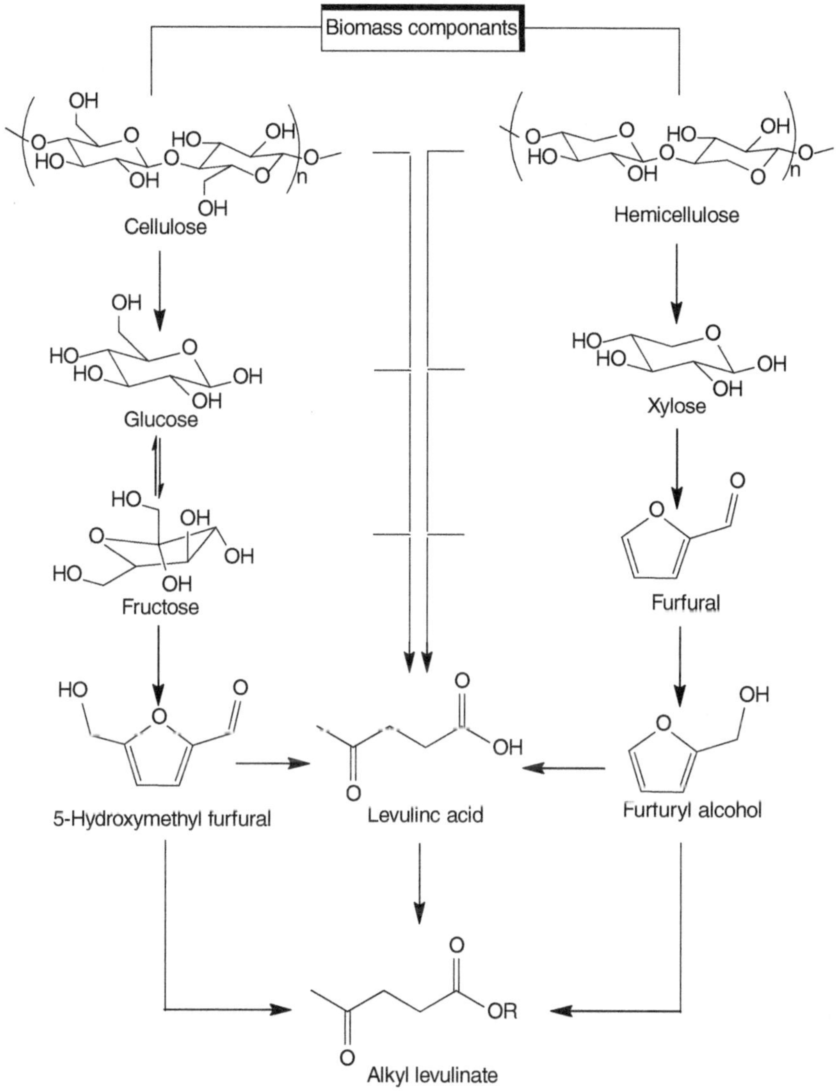

Scheme 3.1 Synthetic routes for alkyl levulinate esters from cellulose and hemicellulose.

reactive acyl cation (Figure 3.3), which is not directly stabilized by the zeolite framework but occupies less space inside the pores, being an alternative route for more constrained zeolites [6].

Other materials that demonstrate interesting results for the esterification of levulinic acid were sulfated metal oxides. For instance, sulfated tin and sulfated

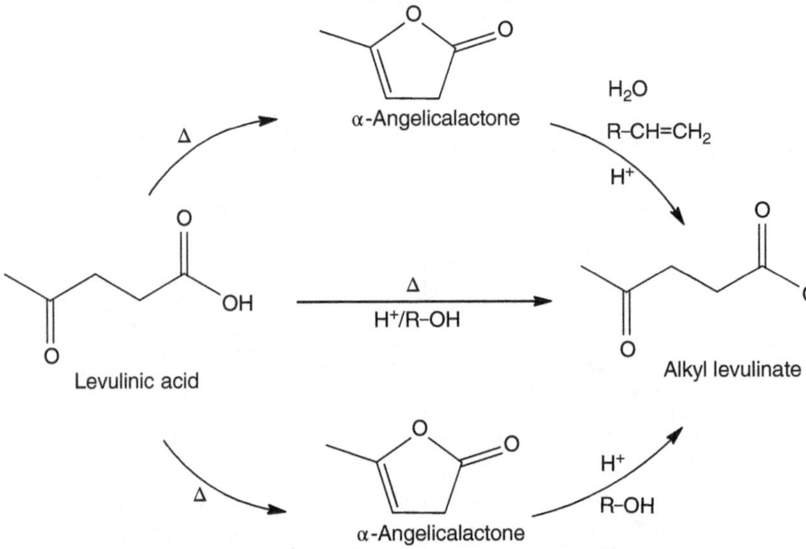

Scheme 3.2 Esterification routes of levulinic acid to alkyl levulinate esters. Source: Kamm et al. [3]/Elsevier.

zirconia oxides presented 77 and 80% yield of AL, respectively [14]. The data showed that morphology, chemical structure, and catalytic activity of the obtained sulfated metal oxide-based materials strongly depend on the synthesis conditions and the structural characteristics of the precursors [14].

Heteropoly acids (HPAs) with Keggin type structures (i.e. $H_3PW_{12}O_{40}$) have also received great attention, especially due to their simple preparation and strong Brønsted acidity. However, these HPAs present disadvantages, such as low-surface areas ($1–10\,m^2{\cdot}g^{-1}$) and problems of separation from the reaction mixture. To improve the efficiency and stability, the HPAs have been supported over different types of materials, such as silica, activated carbon, SBA-15, zeolite, and acid-treated clay [16–19]. The results of levulinic acid esterification on HPAs are highly promising, with yields of AL esters above 90%. However, to achieve such yields, large amounts of catalysts or longer reaction times are required. To circumvent these issues, the silica support can be replaced by ZrO_2, which promotes an increase in the Brønsted acidity of the Keggin units due to the strong W—O—Zr covalent bond between the $H_3PW_{12}O_{40}$ clusters and the ZrO_2 support [19]. Another type of organic–inorganic hybrid material, containing the Keggin units and hydrophobic alkyl groups introduced into the ZrO_2 framework, presented excellent yields for the production of organic levulinates under mild reaction conditions. However, the preparation of these materials is laborious [20].

Table 3.1 Esterification of levulinic acid with alcohols in the presence of heterogeneous acid catalysts.

Catalyst	m_{cat}	Alcohol	Alcohol to acid molar ratio	T (°C)	t (h)	Yield (%)	References
Amberlyst-15	2.5 wt%	Et	5 : 1	70	5	54.0	[6]
Amberlyst-15	30 wt%	Me	20 : 1	66.7	5	82.2	[7]
Amberlyst-15	30 wt%	Et	20 : 1	78.4	5	70.9	[7]
Amberlyst-15	30 wt%	n-Bu	20 : 1	117.7	5	55.3	[7]
Dowex 50WX2	18.9 wt%	n-Bu	3 : 1	80	8	99.5	[8]
Aquivion mP98	10 mol	n-Am	10 : 1	90	24	92.0	[9]
HMCM-22	2.5 wt%	Et	5 : 1	70	5	12.0	[6]
HUSY	2.5 wt%	Et	5 : 1	70	5	7.8	[6]
HBeta	2.5 wt%	Et	5 : 1	70	5	4.2	[6]
HZSM-5	2.5 wt%	Et	5 : 1	70	5	2.9	[6]
HBeta	10 wt%	n-Bu	7 : 1	120	4	82.2	[10]
HMordenita	10 wt%	n-Bu	7 : 1	120	4	29.5	[10]
H-Y	10 wt%	n-Bu	7 : 1	120	4	32.2	[10]
HZSM-5	10 wt%	n-Bu	7 : 1	120	4	30.6	[10]
H-Beta	20 wt%	Et	6 : 1	78	5	39.0	[11]
SBA-15-$(CH_2)_3$-SO_3H	7 wt%	Me	5 : 1	117	2–4	95.0	[12]
SBA-15-$(CH_2)_3$-SO_3H	7 wt%	Et	5 : 1	117	2–4	95.0	[12]
SBA-15-$(CH_2)_3$-SO_3H	7 wt%	i-Pr	5 : 1	117	2–4	90.0	[12]
ZrO_2-SBA-15-SO_3H	5 wt%	Et	10 : 1	70	24	80.0	[13]
SO_4/SnO_2	2.5 wt%	Et	5 : 1	70	5	44.0	[6]
SO_4/TiO_2	2.5 wt%	Et	5 : 1	70	5	40.0	[6]
SO_4/Nb_2O_5	2.5 wt%	Et	5 : 1	70	5	14.0	[6]
SO_4/ZrO_2	2.5 wt%	Et	5 : 1	70	5	9.0	[6]
SnO_2	2.5 wt%	Et	5 : 1	70	5	7.0	[6]
SO_4^{2-}/SnO_2	2.5 wt%	Et	5 : 1	70	7	77.0	[14]
SO_4^{2-}/ZrO_2	2.5 wt%	Et	5 : 1	70	7	80.0	[14]
Sn_2STA/Ta_2O_5	5 wt%	Et	5 : 1	70	3	78.0	[15]
Sn_2STA/Nb_2O_5	5 wt%	Et	5 : 1	70	3	33.0	[15]
Sn_2STA/Al_2O_3	5 wt%	Et	5 : 1	70	3	23.0	[15]
$H_3PW_{12}O_{40}$/K10	10 wt%	n-Bu	6 : 1	120	4	97.0	[16]
$H_3PW_{12}O_{40}$/ZSM-5	20 wt%	Et	8 : 1	78	4	94.0	[17]

(continued)

Table 3.1 (Continued)

Catalyst	m_{cat}	Alcohol	Alcohol to acid molar ratio	T (°C)	t (h)	Yield (%)	References
$H_3PW_{12}O_{40}$/MCM-41	7 wt%	n-Hex	5 : 1	100	10	100	[18]
$H_3PW_{12}O_{40}/ZrO_2$	—	Et	5 : 1	150	3	99.0	[19]
$H_3PW_{12}O_{40}/ZrO_2$-Si(Ph)Si	2 wt%	Me	7 : 1	65	3	99.9	[20]
$H_3PW_{12}O_{40}/ZrO_2$-Si(Ph)Si	2 wt%	Et	7 : 1	78	3	91.5	[20]
$H_3PW_{12}O_{40}/ZrO_2$-Si(Ph)Si	2 wt%	n-Bu	7 : 1	120	3	82.8	[20]
BioC-S3[a)]	25 mg	Et	50 mg/5 ml	130	6	92.0	[21]
SCN[b)]	2.5 wt%	Et	5 : 1	70	5	35.0	[22]
UCC-S[c)]	10 wt%	Et	1 : 1	180	4	95.8	[23]
CCIL[d)]	15 wt%	Et	19 : 1	78	6	66.9	[24]
$SeZrO_2$	0.05 g	Et	1 g/4 ml	120	6	62.0	[25]
S-FCM[e)]	0.05 g	Et	1 g/4 ml	120	6	59.0	[25]
NS-FCM	0.05 g	Et	1 g/4 ml	120	6	61.0	[25]

a) BioC-S3 is a $-SO_3H$ biochar material functionalized with 2-(4-chlorosulfonylphenyl)-ethyltrimethoxysilane (CSPTMS).
b) SCN is a sulfonated carbon nanotube.
c) UCC-S is sulfonated carbon cryogel.
d) CCIL is a carbon cryogel synthesized from ionic liquid and furfural mixtures.
e) FCM is a functional carbon material prepared by hydrothermal carbonization of cellulose using ammonium formate and 5-sulfosalicylic acid additives.

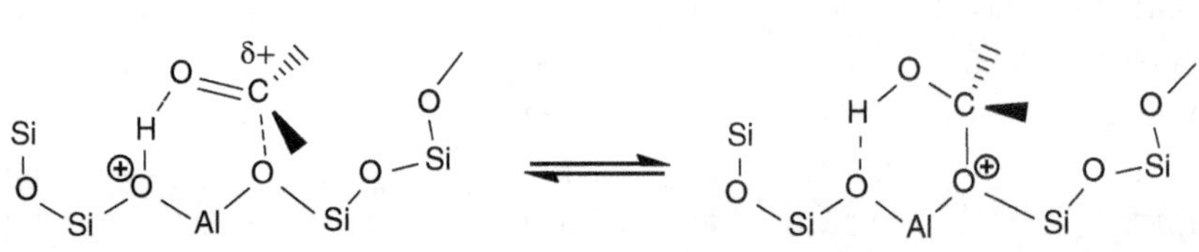

Figure 3.2 Proposed structure for the protonation of the carboxyl group of levulinic acid by the zeolite acid site, with assistance from the framework oxygen atom [6].

Figure 3.3 Proposed structure for the protonation of the hydroxyl group of levulinic acid on the zeolite surface and dehydration to the respective acyl cation (no zeolite assistance) [6].

Other acid catalysts successfully employed in the esterification of levulinic acid include sulfonated materials, such as biochar [21], carbon nanotubes [22], and carbon cryogel [23]. Among these catalysts, sulfonated biochar and sulfonated cryogel presented the best yields of AL esters.

Another esterification route of LA to ALs involves the formation of α-angelicalactone as intermediate, which can afford the levulinate esters through an acid-catalyzed reaction with alcohols [3] or olefins [26] (Scheme 3.3).

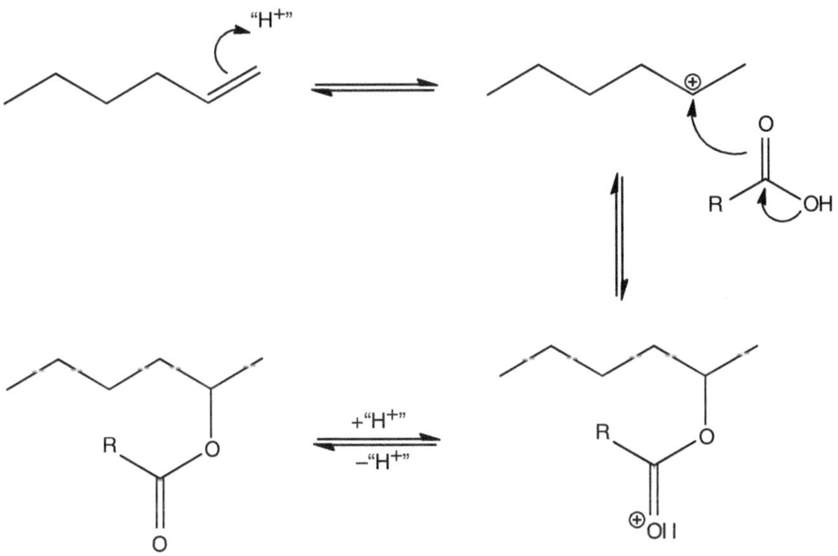

Scheme 3.3 Mechanistic pathway of the acid-catalyzed esterification of the carboxylic acid with olefins.

The use of olefins is more sustainable when compared to the alcoholic pathway because it shows greater atomic efficiency, and no generation of by-products. However, olefin isomerization can simultaneously occur, resulting in mixtures of olefins and, consequently, different AL esters, because olefin isomerization is faster than esterification.

Regarding the mechanistic scheme, it has been reported that esterification of carboxylic acids with olefins proceeds in two steps: first, there occurs protonation of the olefin to form a secondary carbenium ion, which is nucleophilically attacked by the acid [27] (Scheme 3.3).

3.1.2 Direct Production from the Alcoholysis of Polyschacarides

The one-pot production of levulinate ester derivatives from biomass is possible through the alcoholysis of sugars in the presence of acid catalysts. Nevertheless, the process normally requires harsh reaction conditions, such as high temperature (175 °C) and long reaction times (20 hours). There is an increasing interest in this more sustainable strategy, and the key to success is, without doubt, the development of active and selective catalysts. This is a cascade reaction process, where the first step may be the hydrolysis of the polysaccharides to produce hexoses, if the feedstock is composed of cellulosic or hemicellulosic materials. Then, dehydration of the hexoses to 5-hydroxymethyl furfural (5-HMF) occurs, followed by the conversion of the latter into levulinic acid and AL esters, through reaction with the alcohol present in the medium. This approach is especially efficient when fructose is the feedstock, though lignocellulose materials primarily give glucose as the main hexose. Consequently, those catalysts that efficiently isomerize glucose to fructose may give the best results [28, 29].

In alcoholic media, despite solubilization problems, the hexoses can be directly converted into AL derivatives. For instance, ethyl levulinate can be produced from the ethanolysis of hexoses in the presence of an acid catalyst. Xyloses may also be used as feedstock. In this case, the chemical sequence involves dehydration of xylose to furfural, followed by hydrogenation of the latter into furfuryl alcohol, which serves as a precursor to the synthesis of AL esters. The route involving C_5 sugars is more efficient in terms of atom economy [28]. Table 3.2 shows some selected results of sugars conversion to levulinate derivatives over solid acid catalysts.

The catalytic activity of a series of metal sulfates for the synthesis of butyl levulinate from fructose was studied [31]. Among the metal sulfates, $Fe_2(SO_4)_3$ was the most active for this reaction. This solid catalyst presents Lewis acid sites, from the Fe^{3+} ions, and Brønsted acid sites generated from the alcoholysis of Fe^{3+} ions. The catalyst can effectively promote the hydrolysis of fructose and the alcoholysis of the formed 5-HMF producing high yield of the levulinate ester. The pH of $Fe_2(SO_4)_3$ in n-butanol solution was lower than of other metal sulfates. However, the results did not show a correlation between pH and activity. The results revealed that the type of metal sulfates has a more significant impact on catalyst activity than the acidity of the metal sulfate/n-butanol solution. In addition to fructose, other substrates, such as sucrose, glucose, cellulose, and inulin, were also evaluated. Except for cellulose, all of them showed selectivity greater than 50% in the organic levulinate.

Peng et al. evaluated different types of heterogeneous acid catalysts for the conversion of carbohydrates to methyl levulinate. Among the materials tested, sulfated titanium oxide (SO_4^{2-}/TiO_2) was the most promising and showed remarkably high selectivity and yield of methyl levulinate, together with negligible formation of dimethyl ether. Methyl levulinate was obtained in yields of 43, 33, and 59% from sucrose, glucose, and fructose, respectively, at 200°C, 2 hours, and catalyst loading of 2.5 wt% [32].

Table 3.2 Alcoholysis of sugars to ALs over solid acid catalysts.

Catalyst	substrate	m_{cat}	Alcohol	Alcohol to substrate molar ratio	T (°C)	t (h)	Yield (%)	References
Amberlyst 15[a)]	Fructose	1.7 wt%	n-Bu	79 : 1	120	6	13.4	[30]
Amberlyst 39[a)]	Fructose	1.7 wt%	n-Bu	79 : 1	120	6	25.3	[30]
Dowex 50Wx2[a)]	Fructose	1.7 wt%	n-Bu	79 : 1	120	6	28.4	[30]
$Fe_2(SO_4)_3$	Fructose	$25\,g\,l^{-1}$	n-Bu	$5\,g\,l^{-1}$/40 ml	190	3	62.8	[31]
$Fe_2(SO_4)_3$	Sucrose	$25\,g\,l^{-1}$	n-Bu	$5\,g\,l^{-1}$ /40 ml	190	3	50.1	[31]
$Fe_2(SO_4)_3$	Glucose	$25\,g\,l^{-1}$	n-Bu	$5\,g\,l^{-1}$/40 ml	190	3	39.6	[31]
$Fe_2(SO_4)_3$	Celullose	$25\,g\,l^{-1}$	n-Bu	$5\,g\,l^{-1}$/40 ml	190	3	30.5	[31]
$Fe_2(SO_4)_3$	Inulin	$25\,g\,l^{-1}$	n-Bu	$5\,g\,l^{-1}$/40 ml	190	3	56.6	[31]
$Ti(SO_4)_2$	Fructose	$25\,g\,l^{-1}$	n-Bu	$5\,g\,l^{-1}$/40 ml	190	3	41.1	[31]
SO_4^{2-}/TiO_2	qlucose	2.5 wt%	Me	50 ml/2.5 g	200	2	33.0	[32]
SO_4^{2-}/TiO_2	Sucrose	2.5 wt%	Me	50 ml/2.5 g	200	2	43.0	[32]
SO_4^{2-}/TiO_2	Fructose	2.5 wt%	Me	50 ml/2.5 g	200	2	59.0	[32]
SO_4^{2-}/TiO_2	Celullose	2.5 wt%	Me	50 ml/2.5 g	200	2	10.0	[32]
SO_4^{2-}/TiO_2	Starch	2.5 wt%	Me	50 ml/2.5 g	200	2	30.0	[32]
HUSY 0.2	Glucose	0.1 g	Me	0.37 g/12 g	160	5	32.0	[33]
HUSY 0.2	Fructose	0.1 g	Me	0.37 g/12 g	160	5	40.0	[33]
HUSY 0.2	Mannose	0.1 g	Me	0.37 g/12 g	160	5	21.0	[33]
HUSY 0.2	Sucrose	0.1 g	Me	0.37 g/12 g	160	5	38.0	[33]
HUSY 0.2	Cellobiose	0.1 g	Me	0.37 g/12 g	160	5	20.0	[33]
Sn/Mt[b)]	Glucose	0.15 g	Me	0.3 g/24 g	150	6	32.4	[34]
Sn/Mt[b)]	Glucose	0.15 g	Me	0.3 g/24 g	220	6	59.7	[34]
Sn/Mt[b)]	Fructose	0.15 g	Me	0.3 g/24 g	220	6	65.6	[34]
Sn/Mt[b)]	Sucrose	0.15 g	Me	0.3 g/24 g	220	6	62.0	[34]
Sn/Mt[b)]	Cellubiose	0.15 g	Me	0.3 g/24 g	220	6	55.7	[34]
Sn/Mt[b)]	Cellulose	0.15 g	Me	0.3 g/24 g	220	6	19.4	[34]
Sn/Mt[b)]	Starch	0.15 g	Me	0.3 g/24 g	220	6	45.5	[34]
Sn/Mt[b)]	Inulin	0.15 g	Me	0.3 g/24 g	220	6	55.1	[34]
DPW-CeSiMt[c)]	Fructose	0.4 g	Et	0.5 g/10 ml	160	3	65.0	[35]
DPW-CeSiMt[c)]	Glucose	0.4 g	Et	0.5 g/10 ml	170	4	57.0	[35]
DPW-CeSiMt[c)]	Glucose	0.4 g	Me	0.5 g/10 ml	170	4	39.0	[35]
DPW-CeSiMt[c)]	Glucose	0.4 g	n-Pr	0.5 g/10 ml	170	4	52.0	[35]
DPW-CeSiMt[c)]	Glucose	0.4 g	n-Bu	0.5 g/10 ml	170	4	48.0	[35]
DPW-CeSiMt[c)]	Glucose	0.4 g	i-Pr	0.5 g/10 ml	170	4	23.0	[35]

(*continued*)

Table 3.2 (Continued)

Catalyst	substrate	m_{cat}	Alcohol	Alcohol to substrate molar ratio	T (°C)	t (h)	Yield (%)	References
DPW-CeSiMt[c)]	Glucose	0.4 g	Sec-Bu	0.5 g/10 ml	170	4	18.0	[35]
DPW-CeSiMt[c)]	Sucrose	0.4 g	Et	0.5 g/10 ml	170	4	58.0	[35]
DPW-CeSiMt[c)]	Cellobiose	0.4 g	Et	0.5 g/10 ml	170	4	41.0	[35]
DPW-CeSiMt[c)]	Celullose	0.4 g	Et	0.5 g/10 ml	170	5	13.0	[35]
HDS-3.6[d)]	Fructose	40 mg	Et	5 ml/50 mg	200	3.5	70.3	[36]
HDS-3.6[d)]	Glicose	4 mg	Et	5 ml/10 mg	200	8	25.0	[36]
HDS-3.6[d)]	Inulin	4 mg	Et	5 ml/10 mg	200	6	51.2	[36]
HDS-3.6[d)]	Starch	4 mg	Et	5 ml/10 mg	200	8	18.4	[36]
HDS-3.6[d)]	Cellulose	4 mg	Et	5 ml/10 mg	200	10	12.4	[36]
AlPW[e)]	Glucose	0.4 mmol	Me	2 g/15 g	160	0.5	62.9	[37]
AlPW[e)]	Glucose	0.4 mmol	Me	2 g/15 g	160	0.5	64.4	[37]
AlPW[e)]	Fructose	0.4 mmol	Me	2 g/15 g	160	0.5	69.8	[37]
AlPW[e)]	Sucrose	0.4 mmol	Me	2 g/15 g	160	0.5	65.5	[37]
AlPW[e)]	Celullose	0.4 mmol	Me	2 g/15 g	160	0.5	45.2	[37]
SO_4^{2-}/ZrO_2/Sn-Beta	Glucose	0.2 g	Me	0.37 g/12 g	160	5	49.0	[38]
SO_4^{2-}/ZrO_2/Sn-Beta	Fructose	0.2 g	Me	0.37 g/12 g	160	5	59.0	[38]
SO_4^{2}/ZrO_2/Sn-Beta	Sucrose	0.2 g	Me	0.37 g/12 g	160	5	55	[38]
SO_4^{2-}/ZrO_2/Sn-Beta	Mannose	0.2 g	Me	0.37 g/12 g	160	5	31	[38]
SO_4^{2-}/ZrO_2/Sn-Beta	Cellobiose	0.2 g	Me	0.37 g/12 g	160	5	37	[38]
HSiW/Sn-Beta	Sucrose	0.2 g	Me	0.37 g/15 ml	150	3	67	[39]
HSiW/Sn-Beta	Manose	0.2 g	Me	0.37 g/15 ml	150	3	65	[39]
HSiW/Sn-Beta	Fructose	0.2 g	Me	0.37 g/15 ml	150	3	69	[39]
HSiW/Sn-Beta	Starch	0.2 g	Me	0.37 g/15 ml	159	3	58	[39]
HSiW/Sn-Beta	Celullose	0.2 g	Me	0.37 g/15 ml	160	10	62	[39]
Al-5-SB[f)]	Fructose	50 mg	Et	50 mg/5 ml	190	4	58	[40]
BioC-S3[g)]	D-glucose	25 mg	Et	50 mg/5 ml	130	6	3	[21]
BioC-S3[g)]	D-fructose	25 mg	Et	50 mg/5 ml	130	6	28	[21]

a) Reaction carried out in butanol:water ratio of 1 : 19.
b) Mt stands for montmorillonite.
c) DPW-CeSiM is Ce-Si pillared montmorillonite (Ce-SiM) incorporated with Wells-Dawson tungstophosphoric acid ($H_6P_2W_{18}O_{64}$, DPW).
d) HDS-3.6 is a hyper-cross-linked organic polymer synthesized by self-condensation of monomer α,α′-dichloro-*p*-xylene.
e) AlPW is an aluminum phosphotungstate catalyst ($AlPW_{12}O_{40}$).
f) Al-5-SB is alumina-coated silica SBA-15 catalyst.
g) BioC-S3 is a $-SO_3H$ biochar material functionalized with 2-(4-chlorosulfonylphenyl)-ethyltrimethoxysilane (CSPTMS).

Liu et al. demonstrated the catalytic activity of tin-exchanged montmorillonite, which is a natural clay, toward the conversion of different carbohydrate sources in methyl levulinate. They observed yields around 60%. In addition to methyl levulinate, methyl glycoside, methyl lactate, 5-HMF, methyl formate, and humins were observed during glucose conversion. Two possible mechanisms have been proposed. Pathway 1 involves the isomerization of glucose to fructose, dehydration of fructose to 5-HMF and subsequent etherification to form 5-methoxymethyl furfural, followed by rehydration and addition of methanol to form methyl levulinate. In pathway 2, glucose reacts directly with methanol to form methyl glycoside, which undergoes dehydration to form 5-methoxymethyl furfural. The latter undergoes rehydration and methanol addition to form methyl levulinate. However, the conversion of methyl glycoside to 5-methoxymethyl furfural is slow and requires high temperatures. Both Brønsted and Lewis acid sites are involved in methanolysis of glucose [34].

The one-pot synthesis of ethyl levulinate from carbohydrates was studied over Wells–Dawson heteropolyacid supported on Ce–Si pillared montmorillonite (DPW-CeSiM) [35]. The carbohydrates tested were fructose, glucose, sucrose, cellobiose and cellulose. The yield of ethyl levulinate from fructose and glucose was 65% at 160 °C and 57% at 170 °C, respectively. Besides, the glucose alcoholysis were investigated with different alcohols (ethanol, methanol, n-propanol, i-propanol, and sec-butanol). It was observed that the selectivity to the AL ester decreased with the chain length and branching of the alcohol.

HDS-3.6 is a hypercross-linked organic polymer synthesized by self-condensation of α,α′-dichloro-*p*-xylene, via Friedel–Crafts alkylation. This carbonaceous material with a large surface area and strong Brønsted acidity was synthesized for the direct transformation of carbohydrates, such as fructose, glucose, inulin, starch, and cellulose. The catalyst presented nearly complete conversion with 70.3% yield of ethyl levulinate from fructose ethanolysis at 200 °C for 3.5 hours [36].

Zhang et al. developed a series of phosphotungstates, $MPW_{12}O_{40}$ (MPW, where $M = Al^{3+}$, In^{3+}, Cr^{3+}, Fe^{3+}) with Brønsted and Lewis acidity. Among these catalysts, aluminum phosphotungstate ($AlPW_{12}O_{40}$_AlPW) showed superior activity, affording 45–70% yield of methyl levulinate from various carbohydrates at 160 °C, with an extremely low solvent-to-substrate ratio ($7.5\ g{\cdot}g^{-1}$) for 20–30 minutes. The Brønsted and Lewis acid sites of the MPWs exhibited synergistic effects, which facilitated the transformation of glucose to methyl levulinate at mild temperatures. Product analyses indicated that the introduction of metal ions affected the selectivity and provided insights into the reaction pathway. The formation of the pseudo liquid phase of AlPW improved the accessibility of the active catalyst sites, contributing to its high efficiency [37].

A series of alumina-coated SBA-15 mesoporous silica catalysts were prepared via the wet-impregnation method and used in the conversion of fructose to ethyl levulinate. SBA-15 is an ordered mesoporous material; high concentrations of Lewis and Brønsted acid sites can be obtained by dispersing alumina on the SBA-15 surface. Among the catalysts, Al-5-SB with an Al content of 2.2% showed the best results, with 58% yield of ethyl levulinate [40].

3.1.3 Alcoholysis of Furfural

Levulinate derivatives can be directly obtained from the alcoholysis of biomass sources [6, 31, 35, 41, 42]. These processes may be considered more sustainable and efficient in terms of energy. In this context, furfural arises as a promising feedstock. Furfural is a heterocyclic compound containing a furan ring with an aldehyde functional group (COH). It is obtained from pentoses, such as xylose and arabinose, and can be converted to furfuryl alcohol under catalytic reduction conditions. Pentoses obtained from the hemicellulose part of the biomass are traditionally the main raw material to produce furfural. About 62% of the total furfural produced worldwide (280 000 tons/year) is used for the production of furfuryl alcohol [43]. Furfuryl alcohol may undergo acid-catalyzed alcoholysis to afford an AL derivative (Scheme 3.4) [44].

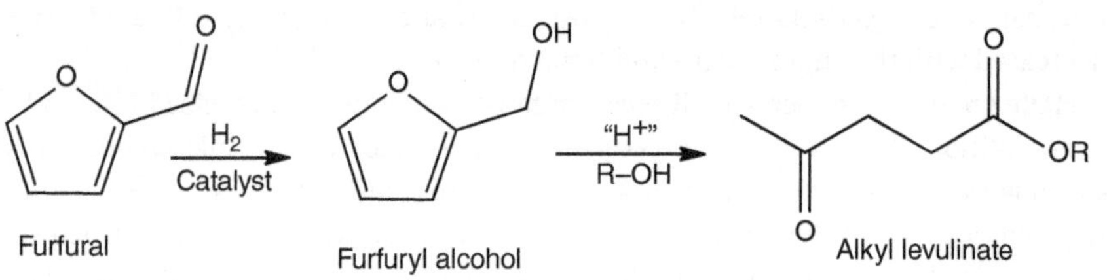

Scheme 3.4 Synthesis of alkyl levulinate esters from furfural.

Table 3.3 shows the direct conversion of furfural to ALs in the presence of bifunctional catalysts. While the metal function catalyzes the hydrogenation of furfural to furfuryl alcohol, the acid support is involved in the alcoholysis of furfuryl alcohol to the correspondent AL.

Typically, these transformations take place under high-pressure conditions using noble metal catalysts and H_2, which is still mainly derived from reforming of nonrenewable natural gas, petroleum, or coal resources [45]. However, the use of catalysts with noble metals makes the process economically unfeasible [49, 50]. An interesting safer, economical, and environmentally friendly approach is the Meerwein–Pondorf–Verley (MPV) reaction under acid catalysis conditions, where an alcohol acts as hydrogen source [46–50]. The direct conversion of furfural to isopropyl levulinate and furan-2-yl-methyl-levulinate was studied over

Table 3.3 Alcoholysis of furfural to ALs in the presence of bifunctional catalysts.

Catalyst	m_{cat}	H_2 pressure	Alcohol	Alcohol to substrate molar ratio	T (°C)	t (h)	Yield (%)	References
$Pt/ZrNbPO_4$	0.2 g	5 MPa	Et	1.5 ml/8.5 ml	130	6	75.7	[45]
Nb_2O_5-ZrO_2	172 mg	—	i-Pr	5 wt%/10 ml	180	8	71.8	[46]
Zr/SBA-15	50 mg	—	Me	100 mg/7 ml	270	10	36.3	[47]
Zr-Al/SBA-15	0.1 g	—	Et	0.2 g/20 ml	180	5	71.4	[48]
Zr-Al/SBA-15	0.1 g	—	i-Pr	0.2 g/20 ml	180	5	19.5	[48]
Zr-Al/SBA-15	0.1 g	—	n-Bu	0.2 g/20 ml	180	5	52.2	[48]
Zr-Al/SBA-15	0.1 g	—	sec-Bu	0.2 g/20 ml	180	5	23.3	[48]
ZrO_2@SBA-15	0.04 g	—	i-Pr	0.3 g/20 ml	160	10	70.0	[49]
Au-HSiW/ZrO_2	0.2 g	—	i-Pr	0.5 mM/20 ml	120	24	80.2	[50]
Au-HSiW/ZrO_2	0.2 g	—	Et	0.5 mM/20 ml	120	24	76.6	[50]
Au-HSiW/ZrO_2	0.2 g	—	n-Pr	0.5 mM/20 ml	120	24	56.1	[50]
Au-HSiW/ZrO_2	0.2 g	—	n-Bu	0.5 mM/20 ml	120	24	50.2	[50]
Au-HSiW/ZrO_2	0.2 g	—	sec-Bu	0.5 mM/20 ml	120	24	62.7	[50]

Nb_2O_5-ZrO_2 mixed-oxide microspheres as bifunctional catalysts for hydrogen transfer hydrogenation and acid-catalyzed alcoholysis in isopropanol. The best catalyst showed selectivity of 71.8% to isopropyl levulinate and 92.6% furfural conversion [46]. Solid catalysts containing Lewis acid sites and neighboring basic sites can promote catalytic hydrogen transfer via the MPV mechanism [49]. Scheme 3.5 shows the direct transfer of hydrogen via six-membered intermediates on a heterogeneous catalyst. The 2-propanol molecule adsorbs on the surface of the catalyst and acts as hydrogen donor, promoting the reduction of the furfural carbonyl group with the formation of furfuryl alcohol.

Mesoporous materials containing zinc has been reported as bifunctional catalysts for the direct production of methyl levulinate from furfural, in quasicritical methanol medium and without the addition of H_2. The Zr-SBA-15 catalyst showed 36.3% yield of methyl levulinate, and 100% furfural conversion at 270 °C for 10 h. No loss of catalyst activity was reported during five recycling of the catalyst [47]. The same SBA-15 material was impregnated with zirconia and alumina for the conversion of furfural to ethyl levulinate. The Lewis and Brønsted acid sites (L/B) ratio were adjusted as a function of the Al_2O_3 and ZrO_2 content on the mesoporous support. Among the evaluated catalysts, Zr-Al/SBA-15 (30 : 10) with an L/B ratio of 2.25 presented the best catalytic performance, with furfural conversion of

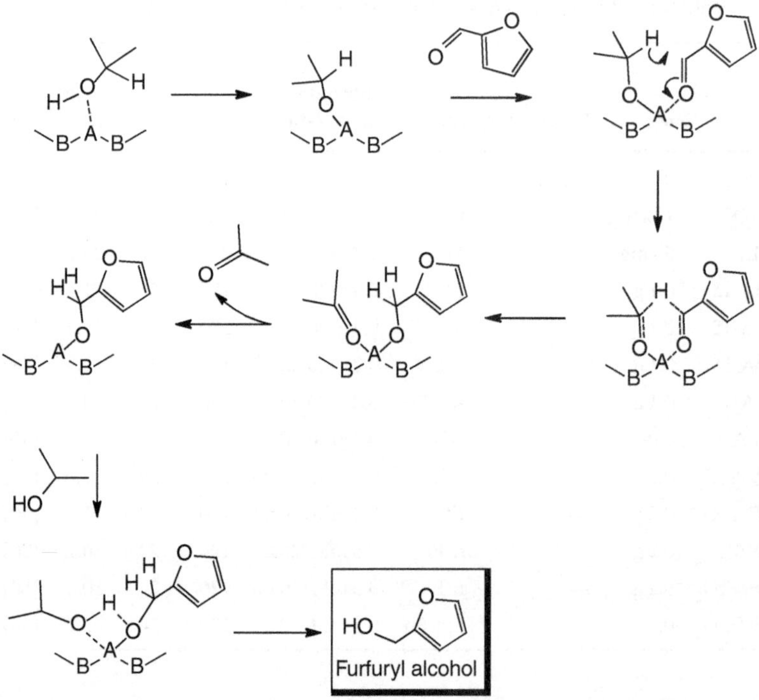

Scheme 3.5 Possible pathway of furfuryl alcohol formation from furfural via MPV reaction over heterogeneous acid catalysts. A = Lewis acid site, B = Lewis basic site. Source: [49]/American Chemical Society.

92.8% and selectivity to ethyl levulinate of 71.4% at 180 °C in 3 h [48]. The direct and highly selective conversion of furfural to furfuryl alcohol and isopropyl levulinate was investigated, using isopropanol as solvent and hydrogen source, over ZrO_2@SBA-15 catalyst. The authors suggested that the furfural MPV reduction to furfuryl alcohol was catalyzed by the Lewis acid sites, whereas the ring-opening etherification of furfuryl alcohol to form the levulinate ester was facilitated by the Brønsted acidity of the catalyst. Furthermore, the SBA-15 supported Zr-based catalyst proved to contain Lewis and Brønsted acid sites [49].

There are some proposed mechanistic pathways of furfuryl alcohol transformation in ALs. Scheme 3.6 presents one of them. Some of the proposed intermediates were identified by NMR or GC-MS analyses [50–52].

According to Scheme 3.6, the electron pair of the hydroxyl group of the furfuryl alcohol interacts with the acid catalyst to yield an oxonium ion. Afterward, the hydroxyl group of the alcohol acts as a nucleophile and an alkoxymethyl furan intermediate is formed. In the sequence, the alcohol might be attached to a furan ring in the presence of the catalyst to form a 1,4-addition product. Further, the

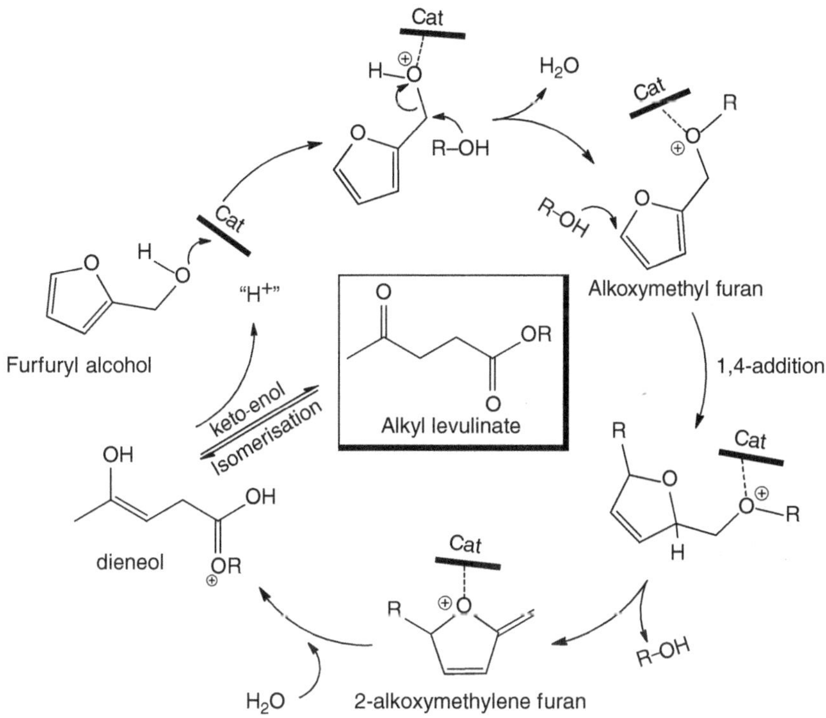

Scheme 3.6 Proposed mechanistic pathway for the acid-catalyzed alcoholysis of furfuryl alcohol to ALs. Source: Rao et al. [51]/Elsevier.

alcohol interacts with the 1,4-addition product to form 2-alkoxymethylene furan, which is further protonated to form the oxonium ion. Finally, the opening of the furan ring occurs to form the dieneol intermediate, which would further be transformed to the AL via keto-enol tautomerization, yet regenerating the catalyst [51].

The conversion of furfuryl alcohol to AL can occur through the isomerization of alkoxy-methyl-furan to AL, or by the reaction of alkoxymethyl furan with alcohol to produce a trialkoxylated ketone as a chemical intermediate, which then undergoes isomerization to form AL. An example of this mechanistic pathway can be observed in Scheme 3.7, where the furfuryl alcohol react with 2-propanol to form isopropyl levulinate [50].

Table 3.4 presents some reported catalysts to convert furfuryl alcohol into AL derivatives, including resins, zeolites, mesoporous aluminosilicates, clays, carbon materials, and others.

Different yields for organic levulinates can be observed from the direct alcoholysis of furfuryl alcohol. The catalysts and reaction conditions, although not completely equivalent, allow us to infer that zeolites have lower selectivity for the

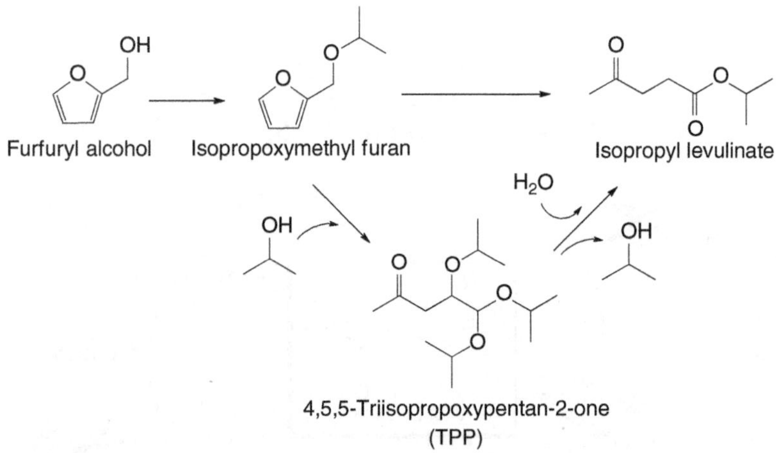

Scheme 3.7 Possible mechanistic pathway of the acid-catalyzed alcoholysis of furfuryl alcohol to isopropyl levulinate. Source: Rao et al. [51]/Elsevier.

Table 3.4 Alcoholysis of furfuryl alcohol to alkyl levulinates in the presence of acid catalysts.

Catalyst	m_{cat}	Alcohol	Substrate to alcohol molar ratio	T (°C)	t (h)	Yield (%)	References
Amberlyst-39	1 g	n-Bu	1/8	100	6	57.0	[53]
OMSA[a)]	60 mg	Et	120 mg/8mL	200	3	80.6	[54]
GO[b)]	0.78 g	Et	0.2 g/3 g	130	2	68.9	[55]
SiO_2	2 wt%	Et	2.5 g/97.5 g	120	4	5.6	[56]
y-Al_2O_3	2 wt%	Et	2.5 g/97.5 g	120	4	13.5	[56]
HZSM-5	2 wt%	Et	2.5 g/97.5 g	120	4	44.4	[56]
HBeta	2 wt%	Et	2.5 g/97.5 g	120	4	60.1	[56]
USY	2 wt%	Et	2.5 g/97.5 g	120	4	67.4	[56]
5%Fe/USY	2 wt%	Et	2.5 g/97.5 g	120	4	73.6	[56]
5%Fe/USY	3 wt%	Et	2.5 g/97.5 g	130	4	90.6	[56]
H-ZSM-5–50[c)]	200 mg	Me	1.6 M FA	170	2	79.0	[57]
H-ZSM-5–50[c)]	200 mg	Et	1.6 M FA	170	2	59.0	[57]
H-ZSM-5–50[c)]	200 mg	n-Pr	1.6 M FA	170	2	60.0	[57]
ZSM-5	3 wt%	n-Bu	1/10	110	6	66.6	[58]
Mordenita	3 wt%	n-Bu	1/10	110	6	4.0	[58]
HBeta	3 wt%	n-Bu	1/10	110	6	10.0	[58]

(continued)

Table 3.4 (Continued)

Catalyst	m_{cat}	Alcohol	Substrate to alcohol molar ratio	T (°C)	t (h)	Yield (%)	References
Al-SBA-15	3 wt%	n-Bu	1/10	110	6	8.0	[58]
HY	3 wt%	n-Bu	1/10	110	6	15.0	[58]
Sapo-34	3 wt%	n-Bu	1/10	110	6	9.0	[58]
Zr/SO_4^{2-}	3 wt%	n-Bu	1/10	110	6	34.0	[58]
Zn_1TPA/Nb_2O_5 d)	0.3 g	n-Bu	98 mg/6 ml	110	5	94.0	[51]
1.0-H_2SO_4/ATTP e)	40 mg	Et	1% FAL in ethanol	160	3	95.4	[59]
Sn_x-TPA/K-10	0.9 mol%	n-Bu	2.3 mmol/45.9 mmol	110	2.5	72.6	[60]
SMWP f)	0.08 g	n-Bu	0.196 g/7.4 g	120	5	90.6	[61]
UCC-S-Fe-300 g)	20 wt%	Et	1 : 1	180	2	95.4	[23]
40Al/DFNS/Pr-SO_3H h)	0.06 g	n-Hex	0.4 ml/3 ml	140	4	93.5	[62]
15WZr800 i)	0.5 g	Bu	8.5/1	190	6	42.8	[63]
BioC-S3 j)	25 mg	Et	50 mg/5 ml	130	6	58.0	[21]
S-FCM k)	0.05 g	Et	0.1g/6 ml	150	6	69.0	[25]
NS-FCM k)	0.05 g	Et	0.1g/6 ml	150	6	62.0	[25]

a) OMSA is an ordered mesoporous SO_4^{2-}/Al_2O_3 solid catalysts.
b) GO stands for graphene oxide.
c) Reaction carried out at 0.2 ml min^{-1} feed rate and 50 bar.
d) TPA is tungstophosphoric acid.
e) ATTP stands for Attapulgite.
f) SMWP is a magnetically recoverable catalyst prepared via a facile methodology involving impregnation, carbonization, and sulfonation
g) UCC is a carbon cryogel
h) DFNS is a dendritic fibrous nanosilica sphere;
i) 15WZr800 is a tungstated zirconia and sulfonated carbon catalysts;
j) BioC-S3 is a $-SO_3H$ biochar functionalized with 2-(4-chlorosulfonylphenyl)-ethyltrimethoxysilane (CSPTMS).
k) FCM is a functional carbon material prepared by hydrothermal carbonization of cellulose using ammonium formate and 5-sulfosalicylic acid additives.

formation of levulinates. However, Zhao et al. evaluated the methanolysis of furfuryl alcohol in continuous flow (0.2 ml·min^{-1}) over zeolite H-ZSM-5 at 170 °C, 50 bar, and high loading of furfuryl alcohol (1.6 M). They found up to 80% yield of methyl levulinate. Angelicalactone was also observed, although in lower amounts when compared to batch reactions. In addition, it was observed that high loadings of furfuryl alcohol produce polyfurfuryl alcohols, which deactivate the catalyst upon pore blockage after one hour of reaction. When 0.2 M of furfuryl alcohol was added, the conversion remained stable for about nine hours, with 71% yield of methyl levulinate [57].

The modification of USY zeolite with Fe was also a strategy to improve the selectivity to organic levulinate from the furfuryl alcohol ethanolysis. According to Hong et al., the presence of Fe in the form of Fe_2O_3 was responsible for the decrease of the number of strong acid sites of the catalyst, which reduced furfuryl alcohol oligomerization and formation of diethyl ether. The yield of methyl levulinate was 90.6% [56].

Besides the mesoporous silicas, such as SBA-15 or MCM-41, there are dendritic fibrous nanosilicas (DFNSs). These materials were tested as support of the heterogeneous acid catalyst (Al/DFNS/Pr-SO_3H) and tested in the reaction of furfuryl alcohol with 1-hexanol. The surface modification of the catalyst was carried out in two steps: direct synthesis of Al/DFNS with different Si/Al molar ratios and synthesis of Al/DFNS/Pr-SO_3H by postgrafting sulfonation. The 40Al/DFNS/Pr-SO_3H catalyst, containing Lewis and Brønsted acid sites, showed the highest activity, with a hexyl levulinate yield and furfuryl alcohol conversion of 93.5% and 99.9%, respectively. The catalyst can be reused for four cycles without significant decrease in product yield [62].

A magnetically recoverable catalyst (SMWP) prepared using waste paper as precursor through an easily methodology, involving impregnation, carbonization, and sulfonation, showed high catalytic activity in the alcoholysis of furfuryl alcohol with 1-butanol. Butyl levulinate was obtained with up to 90.6% yield with furfuryl alcohol conversion around 100% under optimized reaction conditions. Compared to other solid acids, such as HZSM-5 zeolite and Nafion-212 resin, the catalyst showed better performance, which was justified by its strong acidity and good affinity for furfuryl alcohol, resulting from the synergistic effect of SO_3H, COOH, and phenolic OH groups on the carbon surface [61].

Other catalysts that showed excellent results in the conversion of furfuryl alcohol to ALs include zinc-exchanged tungstophosphoric acid supported on niobium oxide, which showed 94% yield of butyl levulinate [51]; tungstophosphoric acid (TPA), partially exchanged with tin, supported on K-10 montmorillonite, which showed 72.6% yield of the same levulinate ester under mild temperature conditions [60]; and acid-treated attapulgite (ATTP), which was prepared upon the treatment with H_2SO_4 and formation of Brønsted and Lewis acid sites [59].

The use of carbon cryogel, produced from urea and furfural and modified via sulfonation and iron doping (UCC-S-Fe), was reported in the ethanolysis of furfuryl alcohol, showing 95.4% yield of ethyl levulinate [23].

3.1.4 Alcoholysis of 5-Hydroxymethyl Furfural

The production of levulinate derivatives from hexoses (glucose, fructose, mannose, among others) occurs in acidic media and involves the formation of

5-HMF as intermediate [2, 64], which may undergo ring cleavage to afford AL and formic acid. Another pathway involves the acid-catalyzed etherification of 5-HMF to 5-ethoxymethyl furfural. This latter compound can also produce AL through ring cleavage (Scheme 3.8) [65].

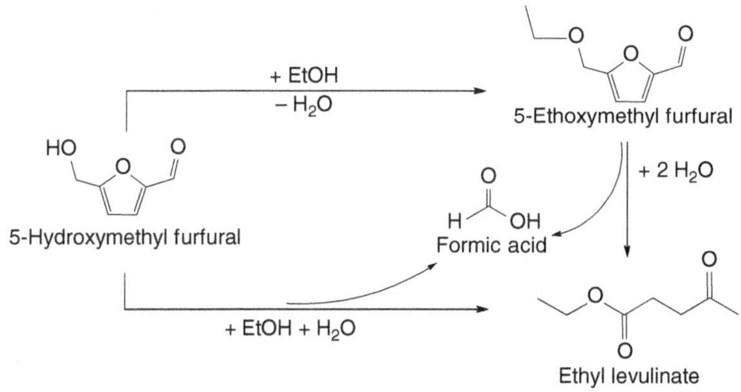

Scheme 3.8 Ethanolysis of 5-HMF to 5-ethoxymethyl furfural and ethyl levulinate.

Alcoholysis of hexoses at high temperatures (120–200 °C) and high alcohol to sugar molar ratios normally produce the correspondent dialkyl ether through dehydration of the alcohol as an undesired side reaction. Thus, the commercialization of ALs using this route may be more difficult. However, some studies indicated that formation of the dialkyl ether is reduced upon the use of low loadings of solid acid catalysts [2]. Another pathway to obtain 5-HMF is the traditional hydroxymethylation of furfural with formaldehyde, using zeolite catalysts under controlled conditions [3, 66].

The use of zeolites to convert 5-HMF into ALs can also generate insoluble organic matter known as humins. It has been reported that in oxygen atmosphere, the formation of humins decreases and the synthesis of methyl levulinate from 5-HMF is accelerated. Since furfuryl methyl ether was observed in the conversion of 5-HMF under O_2 atmosphere, it has been suggested that decarbonylation of 5-HMF may first occur, yielding furfuryl methyl ether, which undergoes methanolysis to form methyl levulinate (Scheme 3.9) [67].

Table 3.5 shows some selected catalytic systems to produce AL esters from 5-HMF.

The cogeneration of humins is the main barrier to the production of ALs from 5-HMF, leading to low overall efficiency and catalyst deactivation. Wang et al. proposed a catalytic system with zeolites and molecular oxygen to solve these problems. Reaction efficiency and catalyst reuse were significantly improved in the presence of 10 bar of oxygen. However, a sharp drop of yield was observed for

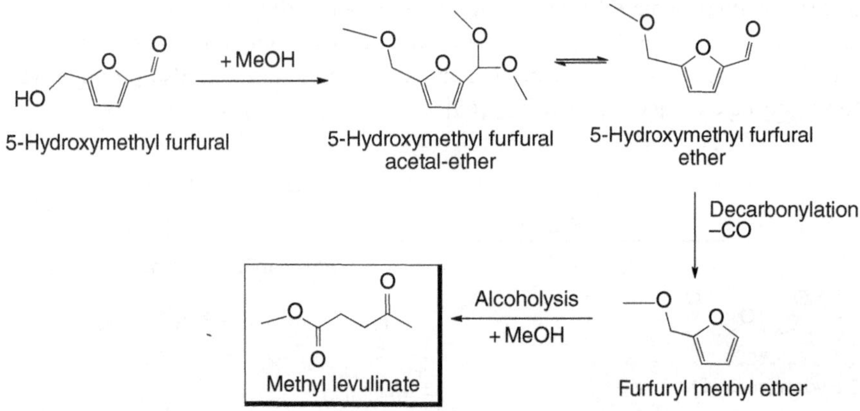

Scheme 3.9 Decarbonylation of 5-hydroxymethyl furfural (HMF) in the methanolysis of 5-HMF in the presence of O_2. Source: Wang et al. [67]/American Chemical Society.

Table 3.5 Alcoholysis of 5-HMF to levulinate esters in the presence of acid catalysts.

Catalyst	m_{cat}	Alcohol	Substrate to alcohol molar ratio	T (°C)	t (h)	Yield (%)	References
MZSM-5[a)]	4.8 wt%	Et	0.085 g/1 g	150	12	90.8	[68]
H-Beta-25	75 mg	Me	1.5 mmol/15 ml	170	0.83	61.5	[67]
H-Beta-40	75 mg	Me	1.5 mmol/15 ml	170	0.83	69.6	[67]
ZSM-5-27	75 mg	Me	1.5 mmol/15 ml	170	0.83	22.6	[67]
ZSM-5-50	75 mg	Me	1.5 mmol/15 ml	170	0.83	20.5	[67]
ZSM-5-130	75 mg	Me	1.5 mmol/15 ml	170	0.83	2.1	[67]
USY-5.4	75 mg	Me	1.5 mmol/15 ml	170	0.83	13.0	[67]
[Cu–BTC][HPM][b)]	40 mg	Et	0.5 mmol/4 ml	100	12	7.6	[69]
[C_3PrIm] [SO_3CF_3]-DFNS[c)]	35 mg	Et	0.27 mmol/ 68.5 mmol	120	3	93.0	[70]
C-SO_3H[d)]	40 mg	Et	0.5 mmol/4 ml	140	8	81.0	[65]
C-SO_3H[d)]	40 mg	Et	0.5 mmol/4 ml	100	6	22.0	[65]
BioC-S3[e)]	25 mg	Me	50 mg/5 ml	130	6	83.0	[21]
BioC-S3[e)]	25 mg	Et	50 mg/5 ml	130	6	84.0	[21]
BioC-S3[e)]	25 mg	n-Pr	50 mg/5 ml	130	6	62.0	[21]
BioC-S3[e)]	25 mg	n-Bu	50 mg/5 ml	130	6	27.0	[21]

a) Hierarchical ZSM-5 zeolite.
b) [Cu–BTC][HPM] is a MOF-based polyoxometalate.
c) [C_3PrIm] [SO_3CF3]-DFNS (C_3 = PrSO_3H) is a Brønsted acid ionic liquid (BAIL) functionalized dendritic fibrous nanosilica sphere catalysts.
d) C-SO_3H is a sulfonic-acid-functionalized carbon nanomaterial, prepared from the direct pyrolysis of MOFs precursor Cu-benzene-1,3,5-tricarboxylate (Cu-BTC) and acidification with sulfuric acid.
e) BioC-S3 is a -SO_3H functionalized biochar materials.

methyl levulinate under oxygen pressures of 20 bar. Zeolites used were H-Beta and H-ZSM-5 with different Si/Al ratios, in addition to USY-5.4. The H-Beta catalysts were the most active, with H-Beta-40 showing 69.6% yield of methyl levulinate, whereas zeolites ZSM-5-130 and USY-5.4 showed poor yields, 2.1% and 13.0%, respectively [67].

Hierarchical ZSM-5 zeolites (MZSM-5) were reported in the synthesis of ethyl levulinate from 5-HMF [68]. This type of zeolites reduces the problems associated with diffusional transportation and enables better access of the active sites to bulky molecules, because of the presence of intra-crystalline mesopores due to the use of a surfactant along with tetrapropylammonium bromide during their synthesis. The selectivity to ethyl levulinate was 90.8%, whereas the standard H-ZSM-5 zeolite presented only 43.7% yield of the product.

The valorization of lignocellulosic agricultural wastes can be directed to the development of heterogeneous catalysts. A series of novel -SO_3H-functionalized biochar materials was produced from vineyard pruning wastes, through hydrothermal treatment with water under subcritical conditions, followed by three different sulfonation processes [21]. 5-HMF conversion and yield of ethyl levulinate were higher than 84% at 130 °C and six hours on the biochar functionalized with 2-(4-chlorosulphonylphenyl) ethyltrimetoxysilane (BioC S3). The results showed that the high acid strength derived from the anchoring of arylsulfonic groups was responsible for the efficient ethanolysis. The BioC-S3 catalyst can be recycled without significant loss of catalytic activity. Authors also found a decrease in the carbon mass balance and darkening of the reaction mixture, especially when 1-propanol and 1-butanol were used. The results were attributed to the formation of high molecular weight compounds through the polymerization of the substrate, suggesting that highly reactive unsaturated intermediates can be formed on the BioC-S3 catalyst surface, leading to heavier products, such as humins.

Carbon nanomaterials functionalized with sulfonic acid groups (C-SO_3H) are active in converting 5-HMF to ethyl levulinate. The nanomaterials were obtained from the direct pyrolysis of the metal-organic framework (MOF) precursor Cu-benzene-1,3,5-tricarboxylate (Cu-BTC), followed by acidification with sulfuric acid. This MOF-assisted approach guarantees a carbon matrix with a large specific surface area that permit the introduction of large number of acid groups. Ethyl levulinate was the main product when the reaction temperature increased from 100 °C to 140 °C, after six hours of reaction. This observation confirms that the furan ring is prone to hydrolytic cleavage to yield the organic levulinate at higher reaction temperatures [65].

ALs can also be prepared from chloromethyl furfural, which can be easily obtained upon the treatment of hexoses with hydrochloric acid. The alcoholysis of the chlorinated derivative produces ALs. The reaction proceeds via dechlorination

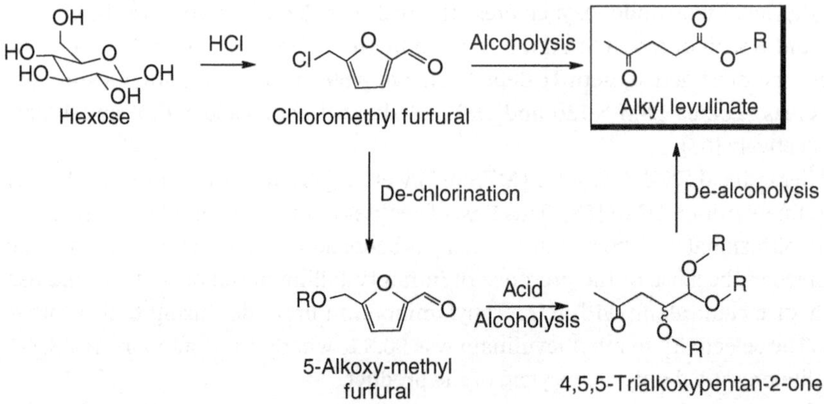

Scheme 3.10 Synthesis of AL from chloromethyl furfural.

to yield 5-alkoxy-methyl-furfural as intermediate at room temperature. On the other hand, at high temperatures, the alcoholysis of chloromethyl furfural produces AL, with formation of 4,5,5-triethoxy-pentan-2-one (TEP) as intermediate [28]. The proposed mechanistic pathway is shown in Scheme 3.10 [52].

3.1.5 Production of Levulinate Inorganic Salts

In the case of inorganic levulinate derivatives, the most traditional production route is the reaction of levulinic acid with metal hydroxides [71] or carbonates [71, 72]. Calcium levulinate, for instance, can be produced by the neutralization of levulinic acid with calcium carbonate, as well as with excess of calcium hydroxide. In the last case, unreacted calcium hydroxide is precipitated as calcium carbonate upon bubbling CO_2 in the reaction medium. Calcium levulinate is precipitated as a di-hydrated salt upon evaporation of the water excess [73].

Similarly, the production of sodium levulinate is not a complicated process and may be produced upon the neutralization of levulinic acid with sodium hydroxide or sodium carbonate. The synthesis of sodium levulinate salt follows an already established methodology that uses 10% solution of levulinic acid heated to 95 °C, with gradual addition of the metal carbonate [72].

The inorganic levulinates have great application potential. The primary use of calcium levulinate, for example, is as calcium supplement, opening great prospects for exploration in the pharmaceutical and food sectors in general. Its application also extends to cosmetics, personal care, and toothpaste.

Sodium levulinate is a substitute for sodium benzoate as preservative in food, beverages and cosmetics [74]. Sodium benzoate is a petroleum-derivatived product, coming from toluene and benzoic acid [75]. Therefore, the market for sodium levulinate in these sectors is extremely advantageous, especially when the use of natural-derived preservatives is considered for the replacement of fossil-derived ones.

3.2 Importance and Market of the Levulinate Derivatives

Organic and inorganic levulinate derivatives are receiving considerable importance in recent years, especially because of their sustainable production from biomass feedstocks. They may find applications in different sectors, such as cosmetics, food, flavors, pharmaceutics, specialty chemicals, and fuels (Figure 3.4). In this latter area, levulinate ester derivatives are considered promising additives for gasoline and diesel fuels, because they improve the viscosity, compressibility, pour and cloud points, as well as lubrication, cleaning, and stabilization of the fuel [26, 28].

Ethyl levulinate is a pale yellow liquid that boils at 206 °C. It is the most important organic levulinate ester, with global sales estimated in US$ 12 million in 2022. The forecast predicts a compound annual growth rate (CAGR) of 3% until 2027. The current major uses are in the food, fragrance and pharmaceutical sectors, but as the commercial production grows and the price goes down, the use in the fuel sector is expected to take an important share of the total market.

Calcium levulinate has been used as a dietary supplement for more than 80 years, whereas sodium levulinate is gaining importance in the cosmetic sector

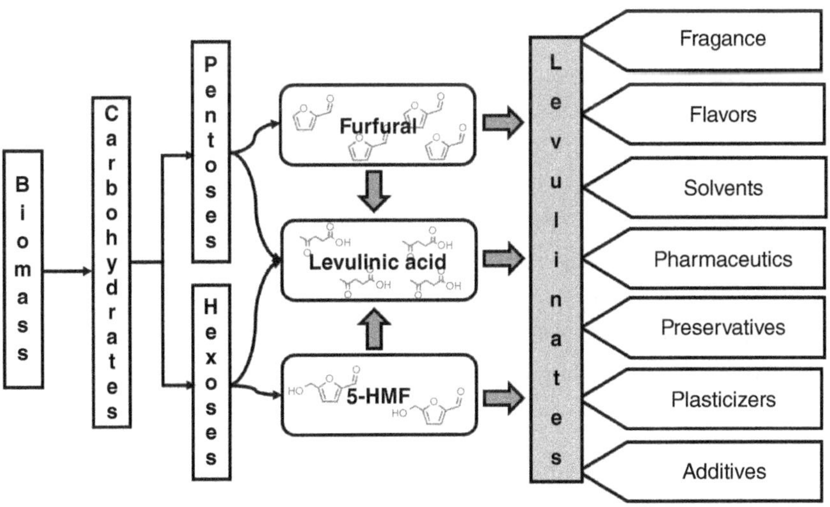

Figure 3.4 Overall utilization of levulinate derivatives.

as a preservative. Both compounds are mostly associated with levulinic acid market share.

3.3 Uses of Organic Levulinate Derivatives

The organic levulinate derivatives are named AL esters (IUPAC name: alkyl 4-oxo-pentanoate) or aryl levulinate esters (IUPAC name: aryl 4-oxo-pentanoate) when the substituent group is an alkyl or an aryl group, respectively. Table 3.6 presents some properties of the most traditional AL esters used in the industry.

3.3.1 Food and Cosmetic

Esters of levulinic acid are especially important in the food and cosmetics sectors as flavoring agents. Ethyl levulinate has a fruity flavor, whereas propyl levulinate has a sweet and caramelized flavoring. Butyl levulinate has herbal and caramel wax flavoring [1, 5, 42, 77, 78]. Ethyl levulinate has been pointed out as an alternative for valencene, which is currently extracted from oranges and used in different beverages. Due to its sweet taste, this product can reduce the amount of sugar in beverages and influence the soft drink industry [42].

Table 3.6 Selected properties of some alkyl levulinate esters.

Property	Ethyl Levulinate	Propyl Levulinate	Butyl Levulinate
Chemical formula	$C_7H_{12}O_3$	$C_8H_{14}O_3$	$C_9H_{16}O_3$
Molecular weight	144.17 g·mol^{-1}	158.19 g·mol^{-1}	172.22 g·mol^{-1}
CAS number	539-88-8	645-76-0	2052-15-5
ChemSpider ID	13853514	191778	15496
Specific mass, g ml^{-1}	1.016 (25 °C)	0.992 (20 °C)	0.974 (25 °C)
Melting Point	<25 °C	<25 °C	<25 °C
Boiling Point	205.8 °C	220.5 °C	237.5 °C
Flash Point	90.56 °C	87.78 °C	92.22 °C
Refractive index	1.420–1.425	1.419–1.425	1.423–1.433
Solubility in water	Soluble	Slightly soluble	Slightly soluble
Solubility in ethanol	Soluble	Soluble	Soluble
Solubility in hydrocarbons	Insoluble	Insoluble	Soluble
Aspect	Colorless to pale yellow liquid	Colorless to pale yellow liquid	Colorless to pale yellow liquid

Source: PubChem [76].

Butyl levulinate can be used as a preservative in the food and cosmetic sectors. The GFBiochemicals' JavelinTM Technology, for example, can produce butyl levulinate from second-generation biomass and the product is a readily biodegradable clear liquid with a low-odor profile. This derivative can replace several chemicals including dibasic esters, glycol ethers, and d-limonene [79].

3.3.2 Fuel Additives

There is an increasing interest in the use of organic levulinate derivatives as fuel additives [6, 28, 48]. Many studies associate levulinic acid production with the development of potential oxygenated additives for transport fuels, with ethyl levulinate being the most quoted for this purpose [28]. However, other levulinate esters have also been reported [50, 61, 80].

The great interest in the levulinate esters lies in the association of their renewable origin and unique properties, such as low toxicity, high lubricity, and flash point. They can also improve cold flow properties, helping to fulfill future needs for gasoline and diesel specifications [28].

Ethyl and butyl levulinates have energy density of 24.8 and 27.1 $MJ\cdot kg^{-1}$, respectively. These values are close to gasoline (31.1 $MJ\cdot kg^{-1}$), diesel (33.6 $MJ\cdot kg^{-1}$), and ethanol (23.5 $MJ\cdot kg^{-1}$). In addition, methyl, ethyl, and butyl levulinates have anti-knocking index values of 106.5, 107.5, and 102.5, respectively [28].

Combustion experiments were carried out in stationary engine and in a constant volume auto-ignition device (IQT). The results of both systems demonstrated that the methyl and ethyl levulinates presented antiknocking quality superior to the reference gasoline [81].

Ethyl levulinate also presents less-negative impact to the environment in case of spillage and leaking to hydric resources, due to its poor water solubility when compared to some traditional fuel additives, such as methyl-*tert*-butyl ether (MTBE) [82]. In addition, the use of ethyl levulinate improves engine efficiency, promotes longer service life, and reduces the emissions of carbon monoxide and nitrogen oxides. This product presents high octane number (107.5) and oxygen content (33 wt%), making it a good candidate as an additive for gasoline compared to other levulinate esters [78]. The main disadvantage is the lower caloric value per kilogram of fuel [83]. As fuel additives for diesel engines, organic levulinates show improvements in fuel combustion, reduction in sulfur emissions, and increase in lubricity [28].

Diesel blends containing 20% of ethyl levulinate have been tested in conventional compression ignition engines. Texaco and Biofine also developed an oxygenated diesel blend that contains 79% of petrodiesel, 20% of ethyl levulinate, and 1% of a co-component [84]. This blend showed cleaner combustion in diesel engines, with a decrease in soot particle emission, as well as good lubrication properties that positively influence the engine's life span. Blends of ethyl levulinate with biodiesel showed improved cold flow properties and reduced solidification point [84]. However, it has been reported that ethyl levulinate tends to form a separate phase in most diesel fuels, whereas butyl levulinate remains completely soluble. The use of butyl levulinate reduces the vapor pressure, improves the lubricity, conductivity, and cold flow properties of the diesel fuel, also leading to cleaner combustion processes with less emission of smoke and NOx. In addition, butyl levulinate is less soluble in water (up to 1.3% by weight) than ethyl levulinate is (up to 15.6% by weight). Both esters exhibit low cetane number, meaning that blending these components in diesel fuels requires the inclusion of some cetane booster component [53].

Ternary mixtures of petrodiesel, biodiesel, and 5 vol% of ethyl levulinate showed good miscibility. The mixtures showed improved closed cup flash point and kinematic viscosity. The CO emission and fuel smoke opacity of the blends were lower than of the pure petrodiesel. However, NOx and CO_2 emissions were higher, and these points might be a barrier to their commercial use [85].

Among the possible candidates as fuel additives, ethyl and butyl levulinates are the most promising. The loading of levulinates in the blends ranges from 0.5 to 20 vol% [53, 86]. There are limitations associated with the solubility of the ALs, particularly at low temperatures. The poor cetane number is also a major concern for diesel blends [87].

3.3.3 Plasticizers

ALs have been considered important intermediates for the production of plasticizing agents [88, 89], surfactants [90], and solvents [89, 91]. Glycerol levulinate ketal (GLK), produced from the reaction of ALs with glycerol, are examples of commercially important plasticizers [88] (Figure 3.5). It has been pointed out the importance of producing rigid polyurethane foams from bio-based platform chemicals, such as glycerol and methyl levulinate [92].

Eugenyl levulinates may act as plasticizers and antibacterial agents for polylactide. It has been shown that the relationship between polar and nonpolar groups in the chemical structure, as well as the molecular weight of the plasticizers, governed the plasticization efficiency and the mechanical properties of the resulting polylactide blends [93]. As the structure of the plasticizer intrinsically determines its performance and, consequently, the plasticizer design, the plasticization of polylactide with isomeric rigid building blocks was studied, combining stereoisomeric cores 1,2-cyclohexanediol and isohexide (isosorbide and isomannide) with levulinic acid (Figure 3.6). The *cis*/*trans*-isomerism seems to be responsible for the better thermal and mechanical properties of the polylactide plasticized with isohexide levulinates, when compared with 1,2-cyclohexanediol

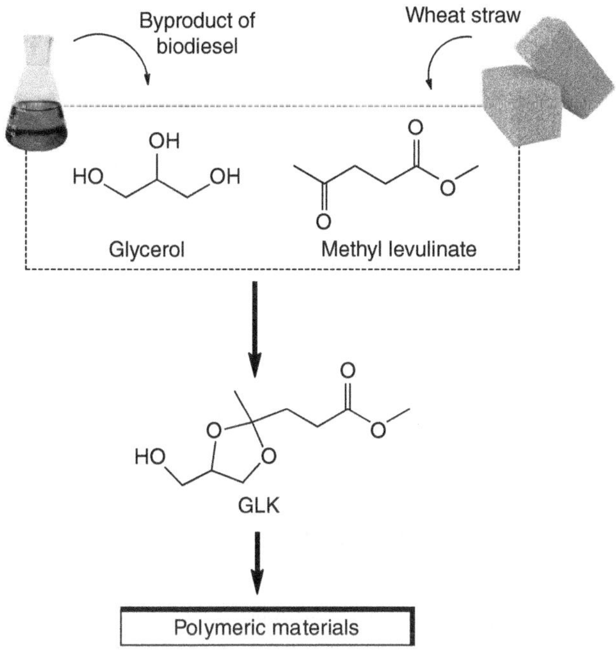

Figure 3.5 Possible application of methyl levulinate in the production of glycerol levulinate ketal (GLK). Source: Mamre Hay/Mamre Produce Pty Ltd.

levulinates. The steric effect has been suggested to be the most important factor to explain these results [94].

3.3.4 Solvents

ALs have been considered green solvents, especially because of some preliminary reports of biodegradation. Therefore, they are being used as green substitutes of traditional industrial solvents [95]. However, recent reports pointed out the toxicity of organic levulinates. The acute lethal concentration (LD_{50}, $mg\,L^{-1}$) of levulinic acid is 23747 ± 1.04, 112.0 ± 1.029 for methyl levulinate, 81.53 ± 1.029 for ethyl levulinate, and 17.04 ± 1.29 for butyl levulinate. As can be observed, levulinic acid is the least toxic compound, and, by contrast, butyl levulinate is the most toxic among the levulinates. The risks to the environment increase with the alkyl chain of the organic levulinates [91, 96].

Levulinate esters are used as biobased solvents (BBS) in polymers, textiles, and coatings. Due to their low health risk, they are also used in the food and pharmaceutical industries [95]. They are also being considered for gas absorption. The solubility of SO_2 in ethyl and butyl levulinates is 5.73 and 4.83 $mol{\cdot}kg^{-1}$, respectively.

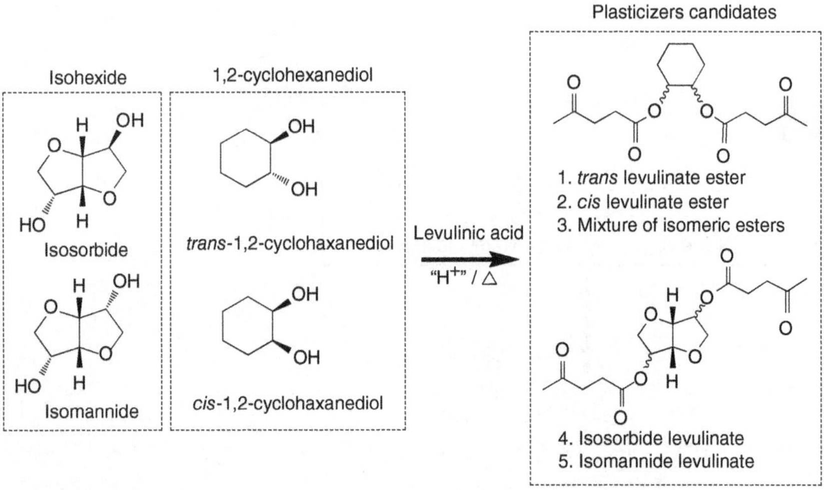

Figure 3.6 Synthetic route for the production of levulinate-derived plasticizers. Source: Xuan et al. [94]/Elsevier.

The dissolution enthalpy was negative in all conditions, and the capturing process was spontaneous. Compared with other absorbents, levulinate BBSs showed potential application for SO_2 removal due to their good capture performance and attractive physical properties [97].

Ethyl and butyl levulinates were also evaluated for CO_2 capture. The CO_2 solubility in butyl levulinate is similar to the solubility in ethyl levulinate. The enthalpy of the solution was negative under all conditions. The solubility of CO_2 in these solvents was further compared with those in ionic liquids (ILs) and ordinary absorbents, indicating that levulinate BBSs presented similar absorption capacity of commercial absorbents, like polyethylene glycol dimethyl ether [98].

Hexylcholinium and octylcholinium levulinates ionic liquids were evaluated as absorbent for volatile organic compounds (VOCs), such as toluene, dichloromethane, and methyl ethyl ketone (MEK). These levulinate-derived solvents were able to absorb VOCs, and no saturation was detected up to concentrations of $3000\,g{\cdot}ml^{-1}$. It is important to mention that classical absorption systems usually involve gas streams with VOC concentrations of up to $1000\,g{\cdot}ml^{-1}$ [99].

Room-temperature ionic liquids (RTILs) and their mixtures with other chemicals have been extensively studied as nonvolatile green solvents for capturing CO_2, SO_2, H_2S, among other gases. Levulinate derivatives have been used in the development of these solvents. For instance, the absorption of carbon dioxide was carried out in methyl-butyl-imidazolium levulinate [C_1C_4Im][LEV] [100]. Levulinate-based RTILs present higher viscosity than similar imidazolium acetate-based ionic liquids, which is a disadvantage for industrial applications in gas absorption processes [101].

The 2-hydroxyethyl-trimethylammonium levulinate ([choline][La]) can be synthesized and mixed with polyethylene glycol (PEG–200) to form a binary mixture used as SO_2 absorbents. The mixture shows higher biodegradability, absorption capacity, as well as lower costs and energy consumption when compared with other ionic liquid systems [102].

A newly explored field is the development of porous ionic liquids, which are considered nonvolatile and versatile. Porous ionic liquids, based on ZIF-8 MOF and phosphonium acetate or levulinate salts, were prepared [103]. The materials showed increased ability to capture carbon dioxide at low pressures. Porous suspensions of phosphonium levulinate ionic liquid reversibly captures 103% more carbon dioxide by mass than pure ZIF-8 at 1 bar and 30 °C. The study demonstrated a new generation of readily available, high-performance liquid materials for effectively capturing carbon at low pressure.

ALs can be used as solvents in organic reactions. For instance, ALs were employed as solvents in the Heck C-C coupling reaction in the presence of heterogeneous Pd-catalysts [104]. Coupling between 2-iodoanisole and *n*-butyl acrylate afforded 88–98% yield of 2-methoxy-cinnamic acid butyl ester in the presence of ALs.

3.4 Uses of Inorganic Levulinate Derivatives

3.4.1 Antifreeze Additive

The most common inorganic levulinate derivatives reported as antifreeze additives are sodium, calcium, and magnesium levulinates [105, 106]. Sodium levulinate was effective at temperatures as low as −12.2 °C. More detailed studies on the environmental impact of these salts are needed, before their large-scale uses as de-icing agents can be commercially considered. Preliminary data demonstrated that their use did not negatively impact the environment, vegetation, or cause corrosivity in vehicles, bridges, and roads.

Sodium levulinate has been reported as a defrosting agent in highways [72], as well as inhibiting or preventing bacterial overgrowth in frozen meats [107] and fresh pork sausage [108].

3.4.2 Cosmetic, Pharmaceutical, and Food

Sodium levulinate is an important component in cosmetic products, improving the antiseptic and stabilizing properties. The cosmetic products are mostly characterized for being of topical or hair utilizations.

Table 3.7 Patents and products containing inorganic levulinate in their formulations.

Origin	Company	Levulinate	Product/uses	References
USA	Johnson & Johnson Consumer	Sodium	Anti aging topical product	[114][a)]
FR	Laboratoires Expanscience	Sodium	Treatment of skin, nails and hairs or mucous membranes	[115][a)]
FR	L'Oréal	Calcium	Treatment of hair fiber	[116]
USA	Colgate Palmolive Co.	Calcium	Slow-release tablet of fluoride and calcium ions	[117][a)]
USA	Welch Foods Inc	Calcium	Source of calcium in beverages	[111][a)]
USA	Stauffer Chemical CO	Calcium	Source of calcium in beverages	[112][a)]
CHI	Universidade Guangxi, da China	Calcium	Source of calcium in beverages	[113]
USA	KV Pharmaceuticals CO	Calcium	Use in vitamins	[118][a)], [119]
CHI	Kaifeng Kangnuo Pharmaceutical Co., Ltd.	Calcium	Use in vitamins	[120]
CHI	Jiangxi Gannan Haixin Pharmaceutical Co Ltd.	Calcium	Use in vitamins	[121][a)]
NZ	Bomac Research Ltd.	Calcium	Use in vitamins	[122][a)]

a) assigned patentes.

Calcium levulinate has been used as a dietary supplement for at least 80 years [73]. In the pharmaceutical industry, for example, this compound can be transformed into tablets, capsules, or injections. In addition, it serves as a nutritional supplement, acting on bone and muscle improvement [109]. Some recent patents were focused on the use of calcium levulinate as food supplement in vitamins [110] and as calcium source in beverages [111–113]. Table 3.7 presents some patents reporting the use of levulinate salts in different product formulations.

3.4.3 Miscellaneous Applications

The North American company Ceramatec Inc. patented a process for the production of sodium levulinate from levulinic acid, and its subsequent use in the synthesis of methyl ethyl ketone (MEK) and octanedione. These ketones are used as solvents in many commercial manufacturing processes, as well as in some household products [123].

Cyclic and aromatic organic compounds, with low oxygen density, can be obtained upon the thermal deoxygenation (TDO) of calcium levulinate at temperatures ranging from 350 to 450°C under inert atmosphere. In this process, there occurs cyclization and dehydration to afford aromatic compounds. Calcium is recovered as $CaCO_3$ and the addition of calcium formate serves as an *in situ* hydrogen source, improving the TDO process. The obtained petroleum-like oil could be further processed in existing refinery facilities [124–127].

The acid-catalyzed hydrothermal degradation of cellulose and the neutralization of the filtrate with calcium hydroxide results in a mixture of calcium levulinate and calcium formate, which upon pyrolysis produces γ-valerolactone, a substance of interest as fuel additive [73, 109, 127, 128].

There are few reports on the use of inorganic levulinates as heterogeneous catalysts. Candu et al. reported the use of sodium levulinate in the synthesis of novel layered double hydroxides (LDH) hybrid catalysts. Sodium levulinate was anchored in the corners of the LDH layers, leading to solid catalysts capable of converting 21.4% of trans-methyl-cinnamate into the corresponding epoxides with 100% selectivity [104].

3.5 Conclusion

Organic and inorganic levulinates are chemicals with great application potential in different areas. Therefore, their world market is expected to increase significantly in the forthcoming years.

For the organic levulinate esters derivatives, the direct route from the alcoholysis of carbohydrates still faces many challenges, especially concerning catalyst development, aiming higher conversion and selectivity at mild conditions. Thus, the traditional esterification of levulinic acid with alcohols appears as the most promising route at short to medium terms. The ALs are promising oxygenated additives for fuels like gasoline and diesel. Hence, the fuel sector may drive the further developments in AL production processes, aiming its large commercial scale utilization.

The inorganic levulinate salts are easily prepared from levulinic acid and find applications in the food, beverage, cosmetic, and pharmaceutical sectors. Calcium levulinate is a traditional calcium supplement, whereas sodium levulinate is gaining importance as preservative.

References

1 Grand View Research (2016). Ethyl levulinate market size, share & trends analysis report by application, by region and segment forecast. https://www.grandviewresearch.com/industry-analysis/ethyl-levulinate-market (accessed October 2021).

2 Démolis, A., Essayem, N., and Rataboul, F. (2014). Synthesis and applications of alkyl levulinates. *ACS Sustainable Chem. Eng.* 2: 1338–1352.

3 Kamm, B., Gerhardt, M., and Dautzenberg, G. (2013). Catalytic processes of lignocellulosic feedstock conversion for production of furfural, levulinic acid, and formic acid-based fuel components. *New Futur. Dev. Catal. Catal. Biomass Convers.* 91–113.

4 Bart, H.J., Reidetschläger, J., Schatka, K., and Lehmann, A. (1994). Kinetics of esterification of succinic anhydride with methanol by homogeneous catalysis. *Int. J. Chem. Kinet.* 26: 1013–1021.

5 Grzesik, M. and Gumula, T. (2001). *Kinetics Models for Esterification of Levulinic Acid with 2-ethylhexanol Using Different Catalysts.* Elsevier Masson SAS.

6 Fernandes, D.R., Rocha, A.S., Mai, E.F. et al. (2012). Levulinic acid esterification with ethanol to ethyl levulinate production over solid acid catalysts. *Appl. Catal. A Gen.* 425–426: 199–204.

7 Ramli, N.A.S., Zaharudin, N.H., and Amin, N.A.S. (2017). Esterification of renewable levulinic acid to levulinate esters using amberlyst-15 as a solid acid catalyst. *J. Teknol.* 1: 137–142.

8 Tejero, M.A., Ramírez, E., Fité, C. et al. (2016). Esterification of levulinic acid with butanol over ion exchange resins. *Appl. Catal., A* 517: 56–66.

9 Trombettoni, V., Bianchi, L., Zupanic, A. et al. (2017). Efficient catalytic upgrading of levulinic acid into alkyl levulinates by resin-supported acids and flow reactors. *Catalysts* 7 (8): 235.

10 Maheria, K.C., Kozinski, J., and Dalai, A. (2013). Esterification of levulinic acid to n-butyl levulinate over various acidic zeolites. *Catal. Lett.* 143 (11): 1220–1225.

11 Patil, C.R., Niphadkar, P.S., Bokade, V.V., and Joshi, P.N. (2014). Esterification of levulinic acid to ethyl levulinate over bimodal micro-mesoporous H/BEA zeolite derivatives. *Catal. Commun.* 43: 188–191.

12 Melero, J.A., Morales, G., Iglesias, J. et al. (2013). Efficient conversion of levulinic acid into alkyl levulinates catalyzed by sulfonic mesostructured silicas. *Appl. Catal., A* 466: 116–122.

13 Kuwahara, Y., Fujitani, T., and Yamashita, H. (2014). Esterification of levulinic acid with ethanol over sulfated mesoporouszirconosilicates: influences

of the preparation conditions on thestructural properties and catalytic performances. *Catal. Today* 237: 18–28.

14 Popova, M., Shestakova, P., Lazarova, H. et al. (2018). Efficient solid acid catalysts based on sulfated tin oxides for liquid phase esterification of levulinic acid with ethanol. *Appl. Catal., A* 560 (April): 119–131.

15 Ganji, P. and Roy, S. (2020). Conversion of levulinic acid to ethyl levulinate using tin modified silicotungstic acid supported on Ta_2O_5. *Catal. Commun.* 134: 105864.

16 Dharne, S. and Bokade, V.V. (2011). Esterification of levulinic acid to n-butyl levulinate over heteropolyacid supported on acid-treated clay. *J. Nat. Gas Chem.* 20 (1): 18–24.

17 Nandiwale, K.Y., Sonar, S.K., Niphadkar, P.S. et al. (2013). *Catalytic Upgrading of Renewable Levulinic Acid to Ethyl Levulinate Biodiesel using Dodecatungstophosphoric Acid Supported on Desilicated H-ZSM-5 as Catalyst.* Elsevier B.V.

18 Wu, M., Zhao, Q.Q., Li, J. et al. (2016). Esterification of levulinic acid into hexyl levulinate over dodecatungstophosphoric acid anchored to Al-MCM-41. *J. Exp. Nanosci.* 11 (17): 1331–1347.

19 Sivasubramaniam, D. and Amin, N.A.S. (2015). Synthesis of ethyl levuli nate from levulinic acid over solid super acid catalyst. *Chem. Eng. Trans.* 45: 907–912.

20 Su, F., Wu, Q., Song, D. et al. (2013). Pore morphology-controlled preparation of ZrO_2-based hybrid catalysts functionalized by both organosilica moieties and Keggin-type heteropoly acid for the synthesis of levulinate esters. *J. Mater. Chem. A* 1 (42): 13209–13221.

21 Peixoto, A.F., Ramos, R., Moreira, M.M. et al. (2021). Production of ethyl levulinate fuel bioadditive from 5-hydroxymethylfurfural over sulfonic acid functionalized biochar catalysts. *Fuel* 303 (April).

22 Oliveira, B.L. and Teixeira Da Silva, V. (2014). Sulfonated carbon nanotubes as catalysts for the conversion of levulinic acid into ethyl levulinate. *Catal. Today* 234: 257–263.

23 Zainol, M.M., Asmadi, M., Iskandar, P. et al. (2021). Ethyl levulinate synthesis from biomass derivative chemicals using iron doped sulfonated carbon cryogel catalyst. *J. Clean. Prod.* 281: 124686.

24 Zainol, M.M., Amin, N.A.S., and Asmadi, M. (2019). Synthesis and characterization of porous microspherical ionic liquid carbon cryogel catalyst for ethyl levulinate production. *Diam. Relat. Mater.* 95 (April): 154–165.

25 Guo, H., Hirosaki, Y., Qi, X., and Lee Smith, R. (2020). Synthesis of ethyl levulinate over amino-sulfonated functional carbon materials. *Renewable Energy* 157: 951–958.

26 Badia, J.H., Ramírez, E., Soto, R. et al. (2021). Optimization and green metrics analysis of the liquid-phase synthesis of sec-butyl levulinate by esterification of levulinic acid with 1-butene over ion-exchange resins. *Fuel Process. Technol.* 220: 106893.

27 Gee, J.C. and Fisher, S. (2015). Direct esterification of olefins: the challenge of mechanism determination in heterogeneous catalysis. *J. Catal.* 331: 13–24.

28 Ahmad, E., Alam, M.I., Pant, K.K., and Haider, M.A. (2016). Catalytic and mechanistic insights into the production of ethyl levulinate from biorenewable feedstocks. *Green Chem.* 18 (18): 4804–4823.

29 Yi, X., Al-Shaal, M.G., Ciptonugroho, W. et al. (2017). Synthesis of butyl levulinate based on A-angelica lactone in the presence of easily separable heteropoly acid catalysts. *ChemSusChem* 10: 1494–1500.

30 Ramírez, E., Bringué, R., Fité, C. et al. (2021). Assessment of ion exchange resins as catalysts for the direct transformation of fructose into butyl levulinate. *Appl. Catal., A* 612: 117988.

31 An, R., Xu, G., Chang, C. et al. (2017). Efficient one-pot synthesis of n-butyl levulinate from carbohydrates catalyzed by $Fe_2(SO_4)_3$. *J. Energy Chem.* 26 (3): 556–563.

32 Peng, L., Lin, L., Li, H., and Yang, Q. (2011). Conversion of carbohydrates biomass into levulinate esters using heterogeneous catalysts. *Appl. Energy* 88 (12): 4590–4596.

33 Zhou, L., Zhao, H., Cui, L. et al. (2015). Promotion effect of mesopore on the conversion of carbohydrates to methyl levulinate over H-USY zeolite. *Catal. Commun.* 71: 74–78.

34 Liu, J., Wang, X.Q., Yang, B.B. et al. (2018). Highly efficient conversion of glucose into methyl levulinate catalyzed by tin-exchanged montmorillonite. *Renewable Energy* 120: 231–240.

35 Lai, F., Yan, F., Wang, P. et al. (2021). Efficient one-pot synthesis of ethyl levulinate from carbohydrates catalyzed by Wells-Dawson heteropolyacid supported on Ce–Si pillared montmorillonite. *J. Clean. Prod.* 324: 129276.

36 Gu, J., Zhang, J., Li, D. et al. (2019). Hyper-cross-linked polymer based carbonaceous materials as efficient catalysts for ethyl levulinate production from carbohydrates. *J. Chem. Technol. Biotechnol.* 94 (10): 3073–3083.

37 Zhang, Y., Chen, X., Lyu, X. et al. (2019). Aluminum phosphotungstate as a promising bifunctional catalyst for biomass carbohydrate transformation to methyl levulinate under mild conditions. *J. Clean. Prod.* 215: 712–720.

38 Jiang, L., Zhou, L., Chao, J. et al. (2018). Direct catalytic conversion of carbohydrates to methyl levulinate: synergy of solid BrØnsted acid and Lewis acid. *Appl. Catal. B*, 220: 589–596.

39 Zhou, S., Yang, X., Zhang, Y. et al. (2019). Efficient conversion of cellulose to methyl levulinate over heteropoly acid promoted by Sn-Beta zeolite. *Cellulose* 26 (17): 9135–9147.

40 Babaei, Z., Najafi Chermahini, A., and Dinari, M. (2018). Alumina-coated mesoporous silica SBA-15 as a solid catalyst for catalytic conversion of fructose into liquid biofuel candidate ethyl levulinate. *Chem. Eng. J.* 352: 45–52.

41 Tian, Y., Zhang, F., Wang, J. et al. (2021). A review on solid acid catalysis for sustainable production of levulinic acid and levulinate esters from biomass derivatives. *Bioresour. Technol.* 342: 125977.

42 Leal Silva, J.F., Grekin, R., Mariano, A.P., and Maciel Filho, R. (2018). Making levulinic acid and ethyl levulinate economically viable: a worldwide technoeconomic and environmental assessment of possible routes. *Energy Technol.* 6: 613–639.

43 Axelsson, L., Franzén, M., Ostwald, M. et al. (2012). Perspective: Jatropha cultivation in southern India: Assessing farmers' experiences. *Biofuels, Bioprod. Biorefin.* 6 (3): 246–256.

44 Kabbour, M. and Luque, R. (2019). *Furfural as a Platform Chemical: From Production to Applications*. Elsevier B.V.

45 Chen, B., Li, F., Huang, Z. et al. (2014). Integrated catalytic process to directly convert furfural to levulinate ester with high selectivity. *ChemSusChem* 7 (1): 202–209.

46 Chen, B., Li, F., Huang, Z., and Yuan, G. (2016). Hydrogen-transfer conversion of furfural into levulinate esters as potential biofuel feedstock. *J. Energy Chem.* 25 (5): 888–894.

47 Chen, H., Ruan, H., Lu, X. et al. (2018). Catalytic conversion of furfural to methyl levulinate in a single-step route over Zr/SBA-15 in near-critical methanol. *Chem. Eng. J.* 333: 434–442.

48 Li, M., Wei, J., Yan, G. et al. (2020). Cascade conversion of furfural to fuel bioadditive ethyl levulinate over bifunctional zirconium-based catalysts. *Renew. Energy* 147: 916–923.

49 Zhang, J., Liu, Y., Yang, S. et al. (2020). Highly selective conversion of furfural to furfural alcohol or levulinate ester in one pot over ZrO_2@SBA-15 and its kinetic behavior. *ACS Sustainable Chem. Eng.* 8 (14): 5584–5594.

50 Zhu, S., Cen, Y., Guo, J. et al. (2016). One-pot conversion of furfural to alkyl levulinate over bifunctional Au-$H_4SiW_{12}O_{40}/ZrO_2$ without external H_2. *Green Chem.* 18 (20): 5667–5675.

51 Rao, B.S., Kumari, P.K., Dhanalakshmi, D., and Lingaiah, N. (2017). Selective conversion of furfuryl alcohol into butyl levulinate over zinc exchanged heteropoly tungstate supported on niobia catalysts. *Mol. Catal.* 427: 80–86.

52 Hovart, J., Klaic, B., Metelko, B., and Sunjic, V. (1985). Mechanism of levulinic acid formation. *Tetrahedron Lett.* 26: 2111–2114.

53 Bringué, R., Ramírez, E., Iborra, M. et al. (2019). Esterification of furfuryl alcohol to butyl levulinate over ion-exchange resins. *Fuel* 257: 116010.
54 Zhang, Z., Yuan, H., Wang, Y., and Ke, Y. (2019). Preparation and characterisation of ordered mesoporous SO_4^{2-}/Al_2O_3 and its catalytic activity in the conversion of furfuryl alcohol to ethyl levulinate. *J. Solid State Chem.* 280.
55 Wu, J., Shao, Y., Jing, G. et al. (2019). Design of graphene oxide by a one-pot synthetic route for catalytic conversion of furfural alcohol to ethyl levulinate. *J. Chem. Technol. Biotechnol.* 94 (10): 3093–3101.
56 Kong, X., Zhang, X., Han, C. et al. (2017). Ethanolysis of biomass based furfuryl alcohol to ethyl levulinate over Fe modified USY catalyst. *Mol. Catal.* 443: 186–192.
57 Zhao, D., Prinsen, P., Wang, Y. et al. (2018). Continuous flow alcoholysis of furfuryl alcohol to alkyl levulinates using zeolites. *ACS Sustain. Chem. Eng.* 6 (5): 6901–6909.
58 Vaishnavi, B.J., Sujith, S., Kulal, N. et al. (2021). Utilization of renewable resources: investigation on role of active sites in zeolite catalyst for transformation of furfuryl alcohol into alkyl levulinate. *Mol. Catal.* 502: 111361.
59 Tian, H., Shao, Y., Liang, C. et al. (2020). Sulfated attapulgite for catalyzing the conversion of furfuryl alcohol to ethyl levulinate: impacts of sulfonation on structural transformation and evolution of acidic sites on the catalyst. *Renew. Energy* 162: 1576–1586.
60 Tiwari, M.S., Dicks, J.S., Keogh, J. et al. (2020). Direct conversion of furfuryl alcohol to butyl levulinate using tin exchanged tungstophosphoric acid catalysts. *Mol. Catal.* 488: 110918.
61 Yang, J., Ao, Z., Wu, H. et al. (2020). Waste paper-derived magnetic carbon composite: a novel eco-friendly solid acid for the synthesis of n-butyl levulinate from furfuryl alcohol. *Renew. Energy* 146: 477–483.
62 Mohammadbagheri, Z. and Najafi Chermahini, A. (2019). Catalytic conversion of furfuryl alcohol to n-hexyl levulinate using modified dendritic fibrous nanosilica. *Chem. Eng. J.* 361: 450–460.
63 Thuppati, U.R., Choi, C., Machida, H., and Norinaga, K. (2021). A comprehensive study on butanolysis of furfuryl alcohol to butyl levulinate using tungstated zirconia and sulfonated carbon catalysts. *Carbon Resour. Convers.* 4: 111–121.
64 Kamm, B., Gruber, P.R., Kamm, M. (2015). Biorefineries - industrial processes and products, in *Ullmann's Encyclopedia of Industrial Chemistry*, pp. 1–38.
65 Wang, Z. and Chen, Q. (2018). Variations of major product derived from conversion of 5-hydroxymethylfurfural over a modified mofs-derived carbon material in response to reaction conditions. *Nanomaterials* 8 (7): 492.

66 Corma Canos, A., Iborra, S., and Velty, A. (2007). Chemical routes for the transformation of biomass into chemicals. *Chem. Rev.* 107 (6): 2411–2502.

67 Wang, Y., Huang, Y., Liu, L. et al. (2020). Molecular oxygen-promoted synthesis of methyl levulinate from 5-hydroxymethylfurfural. *ACS Sustainable Chem. Eng.* 8 (38): 14576–14583.

68 Chithra, P.A. and Darbha, S. (2020). Catalytic conversion of HMF into ethyl levulinate – A biofuel over hierarchical zeolites. *Catal. Commun.* 140: 105998.

69 Wang, Z. and Chen, Q. (2016). Conversion of 5-hydroxymethylfurfural into 5-ethoxymethylfurfural and ethyl levulinate catalyzed by MOF-based heteropolyacid materials. *Green Chem.* 18 (21): 5884–5889.

70 Liu, J., Zhang, C., Song, D. et al. (2021). Efficient transformation of 5-hydroxymethylfurfural to ethyl levulinate over the Brønsted acidic ionic liquid functionalized dendritic fibrous nanosilica spheres. *Microporous Mesoporous Mater.* 326 (May): 111354.

71 Proskouriakoff, A. (1933). Some salts of levulinic acid. *J. Am. Chem. Soc.* 55 (5): 2132–2134.

72 Ganjyal, G., Fang, Q., and Hanna, M.A. (2007). Freezing points and small-scale deicing tests for salts of levulinic acid made from grain sorghum. *Bioresour. Technol.* 98 (15): 2814–2818.

73 Sharath, B.O., Tiwari, R., Mal, S.S., and Dutta, S. (2019). Straightforward synthesis of calcium levulinate from biomass-derived levulinic acid and calcium carbonate in egg-shells. *Mater. Today Proc.* 17: 77–84.

74 Aun, M.V., Mafra, C., Philippi, J.C. et al. (2011). Food additives. *Rev. Bras. Alerg. Imunopatol.* 34 (5): 177–186.

75 Oliveira, P.H.R. and Reis, R.R. (2017). Métodos de Preparação Industrial de Solventes: Ácido Benzóico (CAS 65-85-0). *RVq* 9 (6): 2673–2687.

76 PUBCHEM (2020). Disponível em. https://pubchem.ncbi.nlm.nih.gov/ (accessed March 2020).

77 GFBiochemicals (2015). Levulinic acid is a recognized versatile platform molecule to address a large market potential. http://www.gfbiochemicals.com/products/ (accessed 2021 October).

78 Novita, F.J., Lee, H.Y., and Lee, M. (2017). Energy-efficient design of an ethyl levulinate reactive distillation process via a thermally coupled distillation with external heat integration arrangement. *Ind. Eng. Chem. Res.* 56: 7037–7048.

79 NXTLEVVEL (2022). Application bulletin Antibacterial Activity and Efficacy Antibacterial Activity and Efficacy. https://www.nxtlevvel.com/products/nxt-solv-200/.

80 Srinivasa Rao, B., Krishna Kumari, P., Dhana Lakshmi, D., and Lingaiah, N. (2018). One pot selective transformation of biomass derived chemicals towards alkyl levulinates over titanium exchanged heteropoly tungstate catalysts. *Catal. Today* 309: 269–275.

81 Tian, M., McCormick, R.L., Luecke, J. et al. (2017). Anti-knock quality of sugar derived levulinic esters and cyclic ethers. *Fuel* 202: 414–425.

82 Arteconi, A., Mazzarini, A., and Di Nicola, G. (2011). Emissions from ethers and organic carbonate fuel additives: a review. *Water. Air. Soil Pollut.* 221 (1–4): 405–423.

83 van der Waal, J.C. and de Jong, E. (2016). Avantium chemicals: the high potential for the levulinic product tree. In: *Industrial Biorenewables: A Practical Viewpoint*, 1e (ed. P.D. de María), 97–120. John Wiley & Sons, Inc.

84 Hayes, D.J., Fitzpatrick, S., Hayes, M.H.B., and Ross, J.R.H. (2006). Biofine process – AL, AF et furfural. *Biorefineries - Ind. Process. Prod.* 1: 139–164.

85 Wang, Z., Lei, T., Lin, L. et al. (2017). Comparison of the physical and chemical properties, performance, and emissions of ethyl levulinate-biodiesel-diesel and *n*-butanol-biodiesel-diesel blends. *Energy Fuels* 31 (5): 5055–5062.

86 Wang, Z.W., Lei, T.Z., Liu, L. et al. (2012). Performance investigations of a diesel engine using ethyl levulinate-diesel blends. *BioResources* 7 (4): 5972–5982.

87 Girisuta, B. and Heeres, H.J. (2017). Levulinic acid from biomass: synthesis and applications. In: *Production of Platform Chemicals from Sustainable Resources. Biofuels and Biorefineries* (ed. Z. Fang, R. Smith Jr., and X. Qi), 143–169. Singapore: Springer.

88 Mullen, B.D., Badarinarayana, V., Hall, E.S., Tjossas, M.J., and Leibig, C.M. (2012). Stabilized levulinic ester ketals. WO2013055781A1, Segetis, Inc., issued 18 April 2013.

89 Bloom, P. (2007). Levulinic acid ester derivatives as reactive plasticizers and coalescent solventes. WO2007094922A2, Archer-Daniels-Midland Company, issued 23 August 2007.

90 Harmer, M.A., Kapur, V., and Williams, S.R. (2012). Closed-cell tannin-based foams without formaldehyde. WO2012162645A2, E. I. Du Pont De Nemours And Company, issued 29 November 2012.

91 Lomba, L., Muñiz, S., Pino, M.R. et al. (2014). Ecotoxicity studies of the levulinate ester series. *Ecotoxicology* 23 (8): 1484–1493.

92 Li, P., Xiao, Z., Chang, C. et al. (2020). Efficient synthesis of biobased glycerol levulinate ketal and its application for rigid polyurethane foam production. *Ind. Eng. Chem. Res.* 59 (39): 17520–17528.

93 Xuan, W., Odelius, K., and Hakkarainen, M. (2020). Dual-functioning antibacterial eugenol-derived plasticizers for polylactide. *Biomolecules* 10 (7): 1–16.

94 Xuan, W., Odelius, K., and Hakkarainen, M. (2021). Tunable polylactide plasticizer design: rigid stereoisomers. *Eur. Polym. J.* 157 (May): 110649.

95 Lomba, L., Giner, B., Bandrés, I. et al. (2011). Physicochemical properties of green solvents derived from biomass. *Green Chem.* 13 (8): 2062–2070.

96 Zuriaga, E., Giner, B., Valero, M.S. et al. (2019). QSAR modelling for predicting the toxic effects of traditional and derived biomass solvents on a *Danio rerio* biomodel. *Chemosphere* 227: 480–488.

97 Han, G., Jiang, Y., Deng, D., and Ai, N. (2016). Solubilities and thermodynamic properties of SO_2 in five biobased solvents. *J. Chem. Thermodyn.* 92: 207–213.

98 Deng, D., Han, G., Jiang, Y., and Ai, N. (2015). Solubilities of carbon dioxide in five biobased solvents. *J. Chem. Eng. Data* 60 (1): 104–111.

99 Fahri, F., Bacha, K., Chiki, F.F. et al. (2020). Air pollution: new bio-based ionic liquids absorb both hydrophobic and hydrophilic volatile organic compounds with high efficiency. *Environ. Chem. Lett.* 18 (4): 1403–1411.

100 Yokozeki, A., Shiflett, M.B., Junk, C.P. et al. (2008). Physical and chemical absorptions of carbon dioxide in room-temperature ionic liquids. *J. Phys. Chem. B* 112 (51): 16654–16663.

101 Stevanovic, S., Podgorsek, A., Moura, L. et al. (2013). Absorption of carbon dioxide by ionic liquids with carboxylate anions. *Int. J. Greenh. Gas Control* 17: 78–88.

102 Han, G.Q., Jiang, Y.T., Deng, D.S., and Ai, N. (2015). Absorption of SO_2 by renewable ionic liquid/polyethylene glycol binary mixture and thermodynamic analysis. *RSC Adv.* 5 (107): 87750–87757.

103 Avila, J., Lepre, L.F., Santini, C.C. et al. (2021). High-performance porous ionic liquids for low-pressure CO_2 capture. *Angew. Chem. Int. Ed.* 60 (23): 12876–12882.

104 Candu, N., Paul, D., Marcu, I.C. et al. (2018). Levulinate-intercalated LDH: a potential heterogeneous organocatalyst for the green epoxidation of α,β-unsaturated esters. *Catal. Today* 306: 154–165.

105 Mukherjee, A., Dumont, M.J., and Raghavan, V. (2015). Review: sustainable production of hydroxymethylfurfural and levulinic acid: challenges and opportunities. *Biomass Bioenergy* 72: 143–183.

106 Szklarek, S., Górecka, A., and Wojtal-Frankiewicz, A. (2022). The effects of road salt on freshwater ecosystems and solutions for mitigating chloride pollution – A review. *Sci. Total Environ.* 805: 150289.

107 Thompson, R.L., Carpenter, C.E., Martini, S., and Broadbent, J.R. (2008). Control of *Listeria* monocytogenes in ready-to-eat meats containing sodium levulinate, sodium lactate, or a combination of sodium lactate and sodium diacetate. *J. Food Sci.* 73: 239–244.

108 Vasavada, M., Carpenter, C.E., Cornforth, D.P., and Ghorpade, V. (2003). Sodium levulinate and sodium lactate effects on microbial growth and stability of fresh pork and turkey sausages. *J. Muscle Foods* 14: 119–129.

109 Zhang, J., Wu, S.B., Li, B., and Zhang, H.D. (2012). Advances in the catalytic production of valuable levulinic acid derivatives. *ChemCatChem* 4: 1230–1237.

110 Xiang, H., Lijian, H., Min, X. et al. (2016). A kind of vitamin D2 with calcium levulinate injection composition and preparation method thereof. CN 104000833B. Jiangxi Gannan Haixin Pharmaceutical Co Ltd., issued 18 May 2016.

111 Livisay, S.A. and Lavoie, J.P. (2006). Calcium-fortified, grape-based products and methods for making them. US 7033630B2. Welch Foods Inc., issued 25 April 2006.

112 Melachouris, N. and Lee, C.R. (1988). Calcium fortified acid beverages. US 4740380A; Stauffer Chemical CO., issued 26 April 1988.

113 Ligao, D., Li, L., Kai, L. et al. (2017). A kind of preparation method of grape flavor sugarcane fruit vinegar functional drinks. CN 106281948A. Guangxi University, issued 04 January 2017.

114 Kaur, S., Southall, M.D., Zivin, R.A., Loy, C.J., and Samaras, S.T. (2014). Topical application of 1-hydroxyl 3,5-bis(4′hydroxylstyryl)benzene. WO 2014004177A2. *Johnson & Johnson Consumer, issued 03 January 2014.*

115 Msika, P., Saunois, A., Baudouin, C., Leclere-Bienfait, S., and Debrock, S. (2016). Extract of the above-ground portions of gynandropsis gynandra or cleome gynandra, and cosmetic, dermatological or pharmaceutical composition including same. US 9370541B2. Laboratoires Expanscience, issued 21 June 2016.

116 Plos G. and Bouchara, A. (2013). Cosmetic composition comprising at least one silicone functionalized with one or more mercapto groups anda t least one hygroscopic salt. WO 2013092779A2. L'Oréal, issued 27 June 2013.

117 Grodberg, M.G. (1989). Sustained release fluorine and calcium composition. US 4861590A. Colgate Palmolive Co., issued 29 August 1989.

118 Paradissis, G.N., Levinson, R.S., Heeter, G., Cuca, R.C., and Vanek, P.P. (1999). Multi-vitamin and mineral supplements for women. US 5869084A. Kv Pharmaceuticals Co., issued 09 February 1999.

119 Paradissis, G.N., Levinson, R.S., Cuca, R.C., and Vanek, P.P. (1999). Multi-vitamin and mineral supplements for women. WO0007463A1. Kv Pharmaceutical Company, issued 17 February 2000.

120 Heping, J., Zheng, W., and Yingqiao, L. (2017). A kind of emulsifying process of vitamin D2 and calcium colloid liquid. CN107519131A. Kaifeng Kangnuo Pharmaceutical Co Ltd., issued 29 December 2017.

121 Xiang, H., Lijian, H., Min, X., Fei, L., Xiaozhen, Y., and Qiu Yan, L.T. (2016). A kind of vitamin D2 with calcium levulinate injection composition and preparation method thereof. CN104000833B. Jiangxi Gannan Haixin Pharmaceutical Co Ltd., issued 18 May 2016.

122 Leech, W.F. (2011). Calcium and vitaminB12 based formulation for the treatment of milk fever in animals. NZ582005A. Bomac Research Ltd., issued 29 July 2011.

123 Karanjikar, M. and Bhavaraju, S. (2014). Decarboxylation of levulinic acid to ketone solvents. US 8853463 B2. Ceramatec Inc., issued 07 October 2014.

124 Serrano-Ruiz, J.C., Pineda, A., Balu, A.M. et al. (2012). Catalytic transformations of biomass-derived acids into advanced biofuels. *Catal. Today* 195: 162–168.

125 Eaton, S.J., Beis, S.H., Karunarathne, S.A. et al. (2015). Hydroprocessing of biorenewable thermal deoxygenation oils. *Energy Fuels* 29 (5): 3224–3232.

126 Gunukula, S., Klein, S.J.W., Pendse, H.P. et al. (2018). Techno-economic analysis of thermal deoxygenation based biorefineries for the coproduction of fuels and chemicals. *Appl. Energy* 214: 16–23.

127 Schwartz, T.J., van Heiningen, A.R.P., and Wheeler, M.C. (2010). Energy densification of levulinic acid by thermal deoxygenation. *Green Chem.* 12: 1353–1356.

128 Amarasekara, A.S., Wiredu, B., and Edwards, D.N. (2015). γ-Valerolactone from pyrolysis of calcium salts of levulinic-formic acid mixtures derived from cellulose. *Biomass Bioenergy* 72: 39–44.

4

Levulinic Acid Hydrogenation

4.1 Levulinic Acid Hydrogenation Products

According to the US Department of Energy (DOE), one of the most promising platform chemicals obtained from lignocellulosic biomass is levulinic acid (LA), which was selected as one of the top 12 platform chemicals according to the DOE revised and extended report of 2010 [1].

LA is classified as a key platform chemical for biorefinery processes due to its wide potential applications and relatively low-cost, as well as available technological routes from lignocellulose biomass [2]. Levulinic acid is an excellent precursor for the production of several industrially important products, such as γ-valerolactone (GVL), 1,4-pentanediol (1,4-PDO), 2-methyl-tetrahydrofuran (2-MTHF), valeric acid (VA), alkyl valerates, among others. Scheme 4.1 shows the main products and intermediates that could be obtained upon LA hydrogenation and ring-opening reactions. Among the possible options for LA conversion into valuable products, catalytic hydrogenation has been considered one of the most important and is gaining increasing interest in recent years. Hydrogenation of LA at elevated temperatures may produce 1,4-PDO [3, 4], 2-MTHF [4], pentanoic acid (PA) [5, 6], and 5-nonanone [5], whereas hydrogenation at mild conditions mainly affords GVL.

4.1.1 γ-Valerolactone (GVL)

GVL is a cyclic ester ($C_5H_8O_2$) that presents low volatility, high boiling, and flash points. These properties have motivated the use of this molecule as a solvent and fuel additive. GVL is a colorless liquid with an herbal smell. It is frequently used as food additive and in fragrances [7]; it is also one of the most important lactones and an isomer of δ-valerolactone. Figure 4.1 shows the number of scientific papers, published since 2012, focusing on LA hydrogenation to GVL, indicating the increasing attention that this subject has been receiving in recent years.

Levulinic Acid: A Sustainable Platform Chemical for Value-Added Products, First Edition.
Claudio J.A. Mota, Ana Lúcia de Lima, Daniella R. Fernandes, and Bianca P. Pinto.

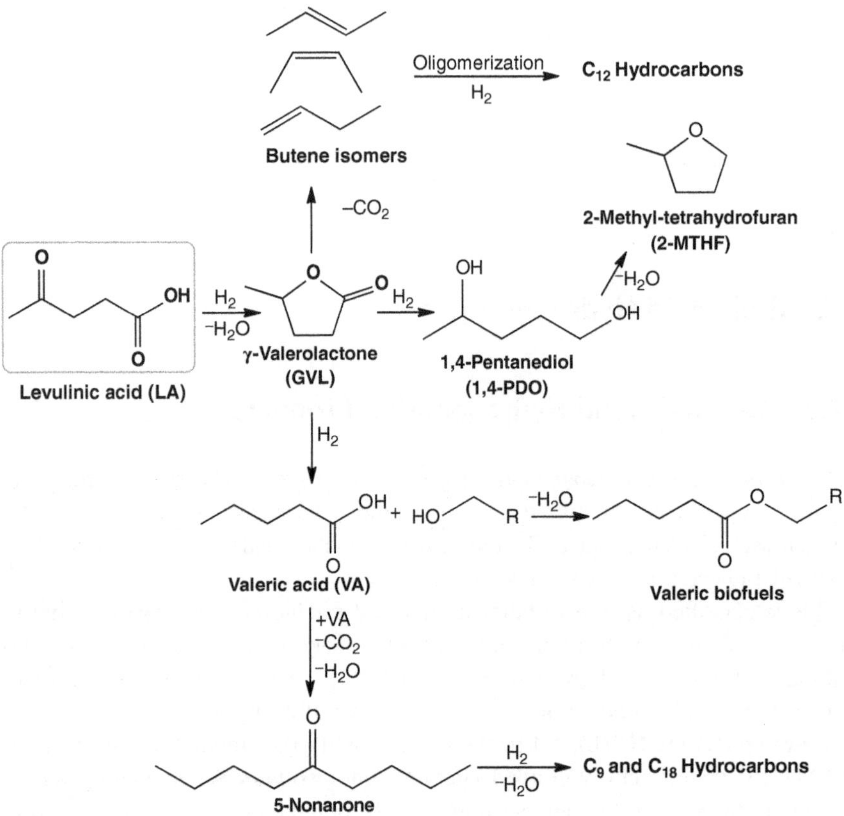

Scheme 4.1 Possible products of catalytic hydrogenation of LA.

GVL may be considered a sustainable chemical. It can be used in the synthesis of polymer building blocks, solvents, and transport fuels [8–10]. Therefore, GVL may be an ideal feedstock for the development of new synthetic pathways to replace petroleum-derived products by biomass-based resources. Some improvements in the synthesis of GVL from biomass-derived sources can be highlighted: (i) the replacement of mineral acids by heterogeneous acid catalysts in the production of LA; (ii) the choice of stable metal catalysts in the presence of mineral acids; and (iii) alternative catalysts to replace noble metal, which would make the system economically more viable [7].

GVL is considered a sustainable precursor for renewable hydrocarbon fuels and can be synthesized using different reaction pathways from cellulose and hemicellulose. The preparation of GVL from cellulose requires hydrolysis and hydrogenation. Another conversion pathway is based on hemicellulose, which is hydrolyzed to form xylose, which is then converted to furfural. In the sequence, furfural is

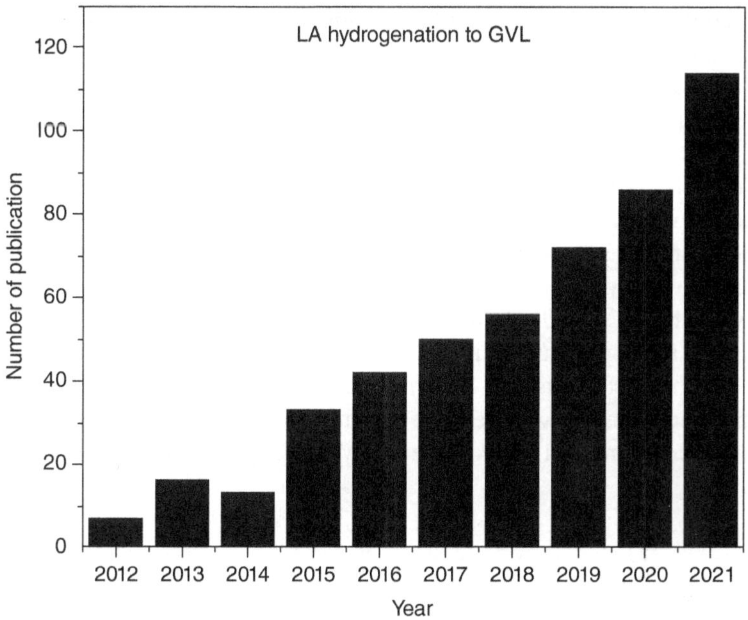

Figure 4.1 Number of published scientific articles on LA hydrogenation to GVL.

converted to furfuryl alcohol upon hydrogenation, and LA may be formed. Finally, a series of hydrogenation–cyclization reactions transforms LA into GVL. In this latter route, two hydrogenation reactions are needed. The lignin fraction can be used as biochar, which can also be used as support for the metal catalysts.

GVL can also be used in the production of adipic acid (1,6-hexanedioic acid), which is an important precursor in the production of polyamides such as polyamide 6,6 (also known as "Nylon") or polyamide 4,6 (referred to as "Stanyl"). Adipic acid esters can be used in plasticizers, lubricants, solvents, and a variety of polyurethane resins. Other uses of adipic acid may include food acidulants, adhesives, insecticides, and tanning agents. The use of GVL in the production of adipic acid is environmentally advantageous, as GVL can be produced from LA, which in turn can be obtained from renewable sources. The reaction pathway involves the initial conversion of GVL to methyl pentenoates in the presence of an acid or a base catalyst. Then, methyl pentenoate is converted to dimethyl adipate through an alkoxy carbonylation reaction. Finally, dimethyl adipate can be converted to adipic acid through simple hydrolysis [11].

The hydrogenation of GVL has stood out due to its versatile and friendly functions [12, 13]. The hydrogenation of LA to GVL can occur either in the vapor or in the liquid phases. Generally, the vapor phase hydrogenation of LA occurs at atmospheric pressure and provides high yields of GVL, but usually requires high-energy input [14]. The liquid-phase hydrogenation of LA is usually carried out under high-hydrogen pressure and in the presence of a solvent [15].

GVL is considered an intermediate in the production of several value-added chemicals, as shown in Scheme 4.2 [7, 16–19]. The hydrogenation of GVL affords 2-MTHF, which is a potential fuel additive. The reaction of GVL and formaldehyde yields α-methylene-γ-valerolactone (MGVL), a new acrylic-like monomer that can be converted into new polymers, because of low toxicity, sustainable origin, and conformational rigidity. This molecule has caught attention due to its similarity with methyl methacrylate and to impart high thermal stability to polymers. MGVL was prepared by heterogeneous gas phase catalytic condensation of formaldehyde with GVL over basic catalysts [20]. The five-membered ring offers some advantages, such as higher refractive index and higher glass transition temperature, in comparison with polymethyl metacrylate (PMMA). Ring-opening of GVL with

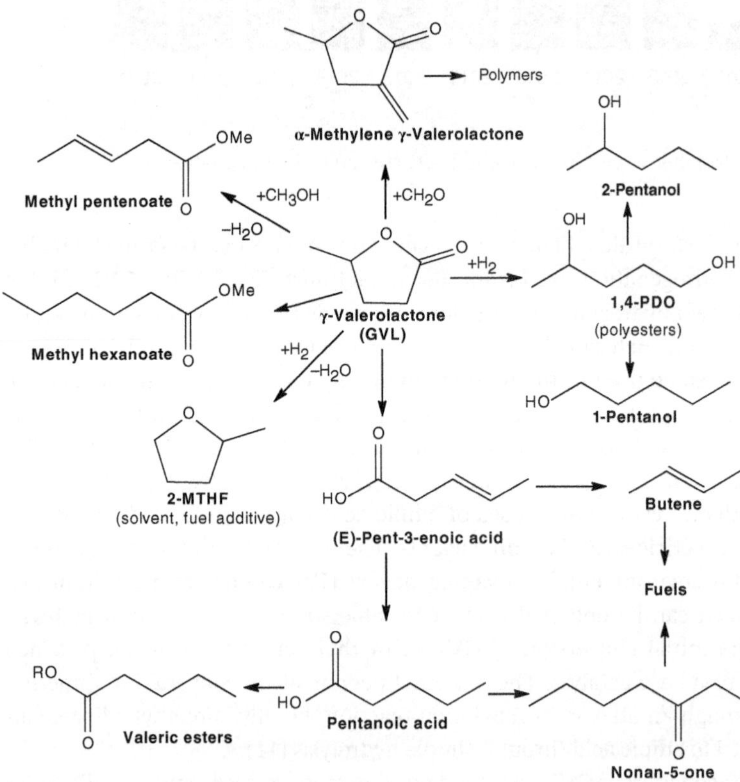

Scheme 4.2 Possible products of GVL conversion.

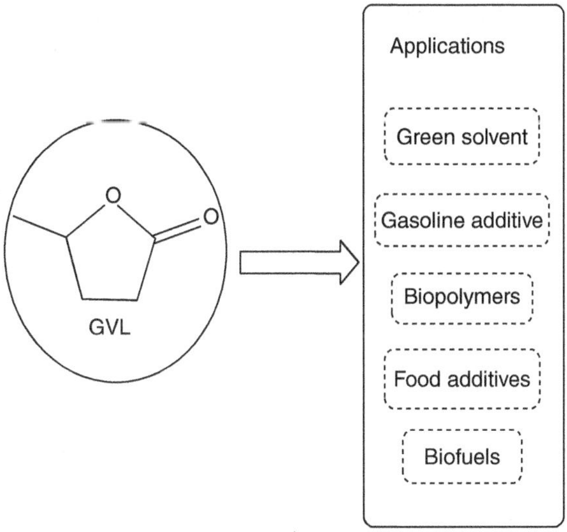

Figure 4.2 Some applications of GVL.

methanol, followed by dehydration, yields methyl-pentenoate, which is an important monomer. Figure 4.2 shows some applications for GVL.

4.1.1.1 GVL Versus Ethanol

GVL is considered a potential fuel additive and a suitable substitute for ethanol in gasoline blends because it presents higher boiling and flash points, also presenting higher calorific value. Therefore, GVL is a good oxygenated additive for gasoline and diesel oil. Although GVL and ethanol are miscible in water, GVL–water mixtures do not form an azeotrope, and separation by distillation is easier and less energy-intensive when compared to the ethanol-water azeotropic distillation. Table 4.1 summarizes some selected properties of GVL and ethanol. The low-vapor pressure of GVL makes its handling safer. On the other hand, the melting temperature can represent a barrier under severe weather conditions.

4.1.1.2 2-Methyl-tetrahydrofuran (2-MTHF)

2-MTHF is an important derivative of LA. It is a colorless liquid with an ether-like odor and is less dense than water. It has been classified by the US Department of Energy as a component of the P-series fuels [21], which can be used independently or blended with gasoline, without any modification of the engine. 2-MTHF is miscible with gasoline in all proportions and has favorable vapor pressure

Table 4.1 Selected properties of GVL and ethanol.

Physical properties	GVL	Ethanol
Formula	$C_5H_8O_2$	C_2H_6O
Mol wt ($g\,mol^{-1}$)	100.12	46.07
Carbon (w%)	60	52
Hydrogen (w%)	8	13
Oxygen (w%)	32	35
Density ($g\,mL^{-1}$)	1.05	0.789
Flash point (°C)	96	13
Melting point (°C)	−31	−114
Boiling point (°C)	207–208	78
Solubility	Miscible with water; soluble in ethanol, acetone	Miscible with water, ethyl ether, acetone, chloroform; soluble in benzene
Refractive index (n20/D)	1.432	1.3611
Acute toxicity LD50 Oral – Rat ($mg\,kg^{-1}$)	8800	7060
Octane number	130	109

properties. In addition, 2-MTHF can be blended in up to 70% with conventional gasoline fuels [22, 23].

2-MTHF is also considered a green solvent and an attractive fuel, due to the large number of raw materials from which it can be obtained. Among them are cellulose, hemicellulose, C_5, and C_6 sugars, as well as LA and furfural [24, 25]. These intermediates can undergo catalytic reduction to produce 2-MTHF as the desired product.

There are several studies on the synthesis of GVL and 2-MTHF through liquid-phase hydrogenation of LA, using heterogeneous and homogeneous catalysts, between 50 and 200 bar of H_2 pressure [4, 26–29].

4.1.1.3 1,4-Pentanediol (1,4-PDO)

Biogenic diols, especially 1,4-PDO, can be produced from LA through a catalytic pathway and used as monomers for the production of high-strength biodegradable polyesters [30]. 1,4-PDO can also be used as an intermediate in the production of pharmaceuticals, cosmetics, and fine chemicals [4]. Table 4.2 shows some selected properties of 2-MTHF and 1,4-PDO.

Table 4.2 Selected properties of 2-MTHF and 1,4-PDO.

Physical properties	2-MTHF	1,4-PDO
Formula	$C_5H_{10}O$	$C_5H_{12}O_2$
Mol wt (g mol^{-1})	86.13	104.15
Density (g ml^{-1})	0.86	0.986
Melting point (°C)	−136	50.9
Boiling point (°C)	78–80	72–73
Refractive index (n20/D)	1.406	1.447
Solubility in water (mg l^{-1})	150	Soluble

4.1.1.4 Alkyl Valerates

Alkyl valerates are known as valeric biofuels, which are suitable to be mixed with gasoline or diesel, depending on the length of the alkyl chain of the added alcohol. When short-chain alcohols, like methanol and ethanol, are used in the esterification of valeric acid (VA), the obtained methyl and ethyl valerate esters are suitable as gasoline additives. On the other hand, with longer-chain alcohols, such as butanol and pentanol, the respective valerates are more suitable as diesel additives, presenting excellent energy density as well as volatility and ignition properties [31, 32].

VA can also be used as a starting material for decarboxylative coupling reactions, to produce 5-nonanone, which is an interesting starting material for the manufacture of C_9–C_{18} linear and branched hydrocarbons through hydrogenation, dehydration, isomerization, and oligomerization reactions [33].

4.2 Performance of GVL as Fuel Additive

With the progressive abolition of methyl-*tert*-butyl ether (MTBE) as an oxygenated fuel additive to improve the combustion efficiency and the anti-knocking properties, some oxygenated candidates such as ethanol and GVL are being considered for that purpose.

It has been suggested [8] that the performance of GVL as a fuel additive is similar to that of ethanol. Studies of gasoline blends containing up to 10% GVL and 10% ethanol showed practically the same research octane number (RON) and similar physical properties. Compared to other oxygenates, such as methanol, ethanol, MTBE, and ethyl-*tert*-butyl ether (ETBE), GVL presented the lowest vapor pressure, which is important for controlling the emission of volatile organic compounds (VOCs) [32].

It has been reported [34] that addition of 7.1% GVL to diesel-biodiesel blend has little effect on engine performance and NOx emission but reduced the concentration of CO, unburned fuel, and smoke.

GVL has been proposed as an illuminating liquid for burning coal without creating smoke or odor, also producing low VOC emissions [35].

4.3 Levulinic Acid to γ-Valerolactone

In general, GVL can be synthesized by hydrogenation of LA via two reaction pathways: (i) endothermic acid-catalyzed dehydration of LA, above 180 °C, to form α-angelicalactone through intramolecular esterification, which is subsequently hydrogenated to GVL and (ii) hydrogenation of the keto-carbonyl group of LA to 4-hydroxyvaleric acid followed by intramolecular esterification [19] (Schemes 4.3 and 4.4). In both pathways, the hydrogenation step depends on the activity of the metal catalyst, whereas dehydration and ring closure are influenced by the acid function of the medium [37–39].

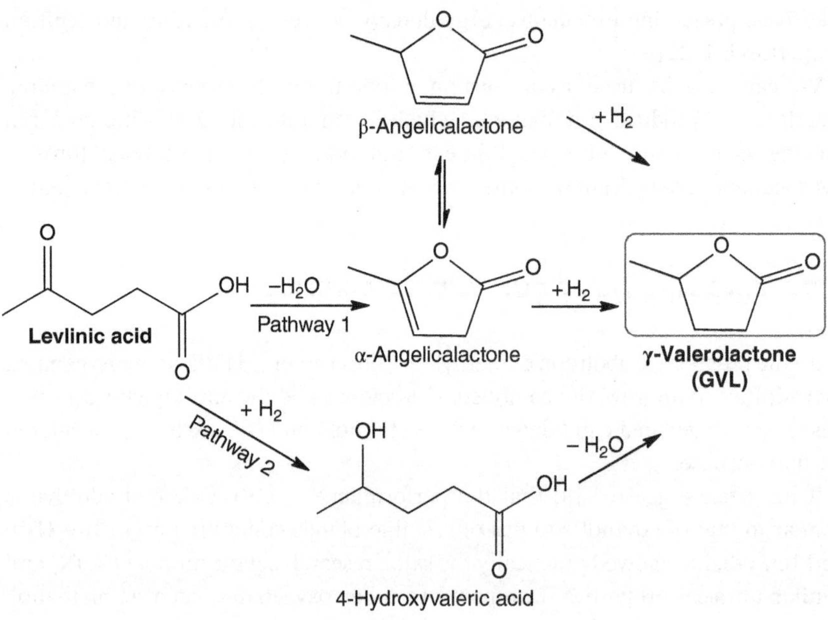

Scheme 4.3 Reaction pathways of LA to GVL.

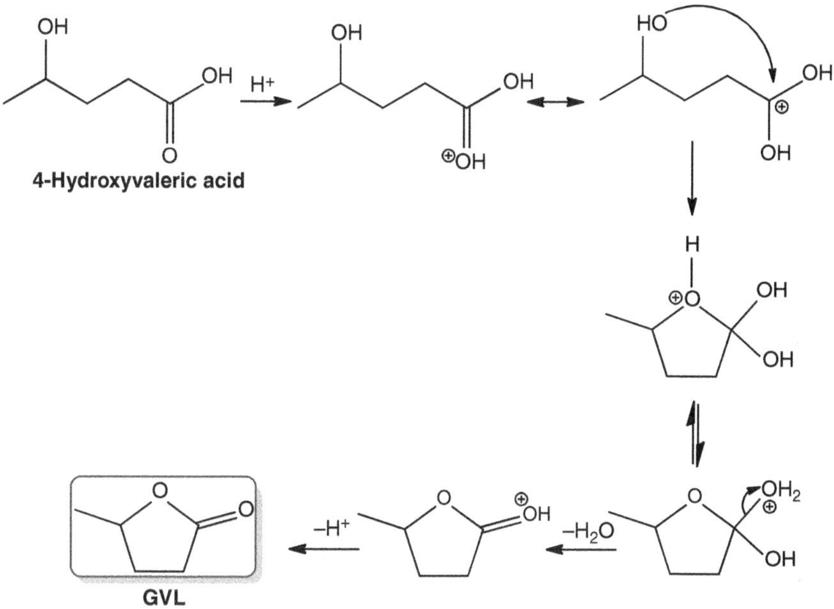

Scheme 4.4 Possible mechanistic pathway for the production of GVL from 4-hydroxyvaleric acid intermediate. Source: Ruppert et al. [36].

At high temperatures and in the presence of a heterogeneous acid catalyst, LA dehydration affords α angelicalactone, which readily produces coke causing deactivation of the heterogeneous catalytic system [10].

4.3.1 Conversion of GVL into 1,4-PDO and 2-MTHF

GVL can be hydrogenated to 1,4-PDO [41]; therefore, the production of 1,4-PDO from LA passes to GVL as intermediate (Scheme 4.5). The selectivity to GVL or 1,4-PDO depends on the specific catalytic performance of the catalysts, comprising hydrogenation, lactonization, and ring-opening of GVL, which is related to the distribution of hydrogenation sites, acid/base sites, and pore characteristics of the catalyst.

The conversion of GVL under solvent-free conditions yields 2-MTHF in 43% selectivity at 190 °C, 100 bar, and 24 hours [42]. Some side reactions have been identified resulting in by-products such as 1,4-PDO, 2-butanol, butane, 2-pentanol, 1-pentanol, pentane, and methane (Scheme 4.6). Investigating the origin of the main reaction products, it was found that 2-MTHF undergoes hydrogenation to 2-pentanol for reaction times longer than 24 hours. The presence of water in the

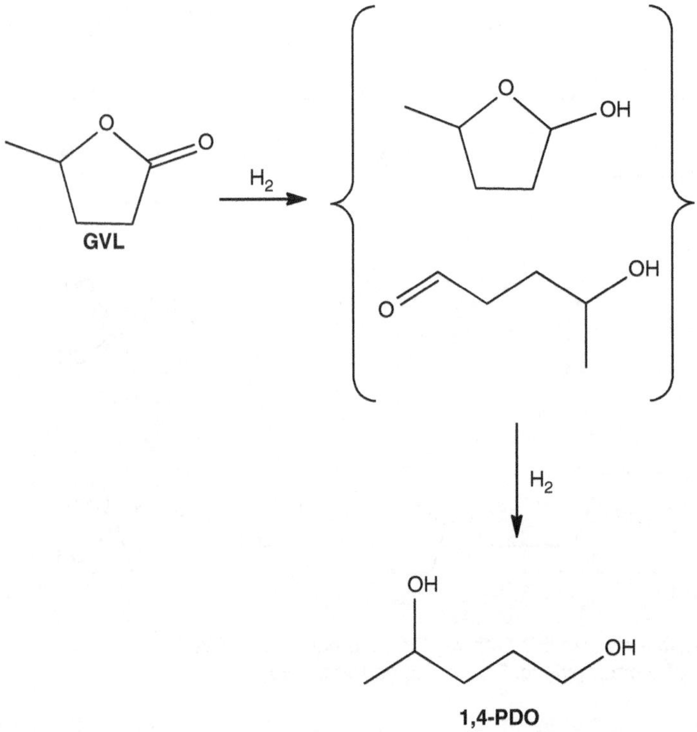

Scheme 4.5 GVL hydrogenation to 1,4-PDO.

system inhibits dehydration and displaces the reaction to form 2-pentanol, instead of the desired production of 2-MTHF. The low yield of 1,4-PDO was attributed to its rapid transformation into 2-MTHF [43], supporting the fact that formation of 1,4-PDO is the rate-determining step in the GVL hydrogenation.

2-MTHF can be converted to 2-pentanol with traces of pentane, 1,4-PDO, and 2-butanol, which suggests that 2-pentanol is formed by catalytic hydrogenation of 2-MTHF [44], as well as from 1,4-PDO hydrogenation. A sequence of dehydration/hydrogenation reactions, starting with 2-pentanol, yields pentane. The presence of 1,4-PDO and 2-butanol confirms the reversibility of the dehydration and dehydrogenation of 1,4-PDO to GVL and 4-hydroxypentanal, respectively.

The direct conversion of LA to 2-MTHF was first reported in 1947 [45] in the presence of $Cu_2Cr_2O_5$. The production of 2-MTHF from GVL involves hydrogenation, whereas the one-pot synthesis from LA, besides hydrogenation, also involves dehydration with the production of water as a by-product [42]. Thus, the formed water can slow down the further dehydration of 1,4-PDO to 2-MTHF. While GVL is hydrogenated to 1,4-PDO, water inhibits the acid-catalyzed dehydration of 1,4-PDO to 2-MTHF. In this way, 1,4-PDO undergoes a subsequent transformation to 2-butanol and 2-pentanol (Scheme 4.7).

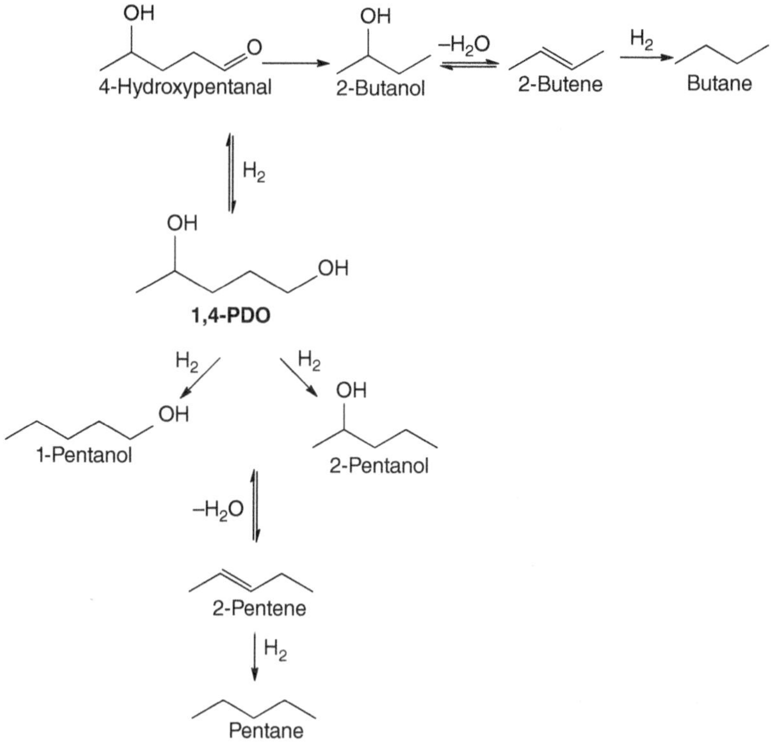

Scheme 4.6 1,4-PDO hydrogenation to other products.

Hydrogenation of GVL at 100 bar and 190 °C, using 5% Ru/C as catalyst for 48 hours, showed traces of 2-MTHF and 1,4-PDO. The main products were 2-pentanol, 2-butanol, and methane, respectively. 2-Pentanol is formed by catalytic hydrogenation of 2-MTHF, as well as from hydrogenation of 1,4-PDO. The dehydration/hydrogenation of 2-pentanol produces pentane [42].

4.3.2 GVL to Butenes and Hydrocarbons

Aqueous solutions of GVL can be converted to pentanoic acid (PA) by the combined ring opening on the acid sites, and hydrogenation over the metallic sites when a bifunctional catalyst, such as Pd/Nb_2O_5, is employed [38].

Production of PA is accompanied by the formation of butane, which is an important raw material for the production of plastics, through decarboxylation of PA over the metallic sites [46], also releasing stoichiometric amounts of CO_x. As a result of successive hydrogenation and dehydration of PA on metallic and acidic sites, *n*-pentane is formed. Pentanoic acid can be converted to 5-nonanone releasing CO_2 and water. When increasing the temperature and decreasing the space velocity, this transformation can be made with the same Pd/Nb_2O_5 catalyst

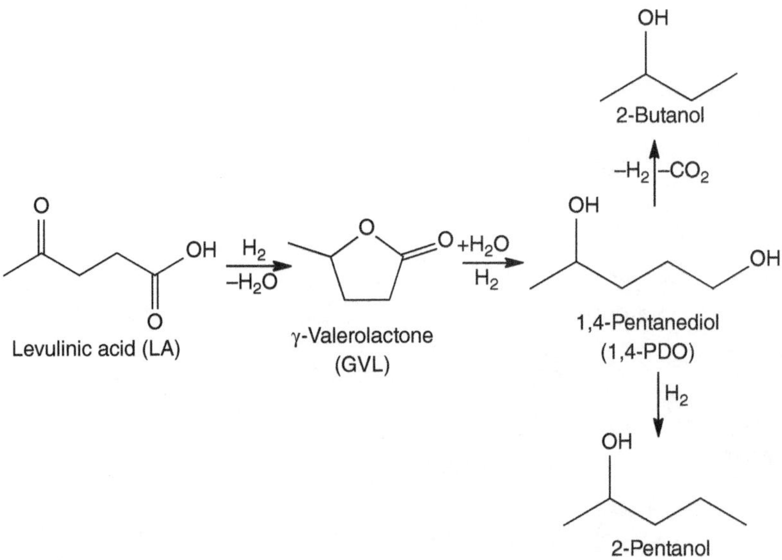

Scheme 4.7 One-pot catalytic hydrogenation of LA to aliphatic alcohols.

used in the production of PA, which allows the direct production of 5-nonanone from aqueous GVL in a single catalytic bed [38]. 5-Nonanone has applications as an industrial solvent for paints and resins and can be used as a precursor to hydrocarbon fuels, such as gasoline and diesel. Through hydrogenation/dehydration cycles, using Pt/Nb_2O_5, 5-nonanone can be converted to linear nonane, which presents good cetane number and lubricity for use as diesel fuel.

Dumesic and coworkers studied the conversion of GVL to liquid alkene hydrocarbons with molecular weight for use in transport fuels, through an integrated catalytic system that does not require an external source of hydrogen. The GVL feed undergoes decarboxylation at 36 bar over silica-alumina catalyst to produce a gas stream containing butene and CO_2. This gaseous stream is fed directly to an oligomerization reactor containing an acid catalyst, such as HZSM-5 or Amberlyst-70, which couples the butene monomers to form liquid alkenes with molecular weights in the gasoline and/or jet fuel range [44] (Scheme 4.8).

The hydrolysis of lignocellulosic biomass affords sugars and lignin that can be processed by biological or chemical means. Biomass hydrolysis is usually complex and offers selective processing options and unavailable platform chemicals compared with thermochemical technologies [47].

The application of GVL as a transport fuel has some limitations, such as low-energy density, mixing limits, and high-water solubility. As GVL is usually produced from LA in aqueous media, the direct use of GVL as fuel requires purification that normally involves distillation and extraction methods to remove water, leading to increased process costs. The use of solvents such as ethyl acetate

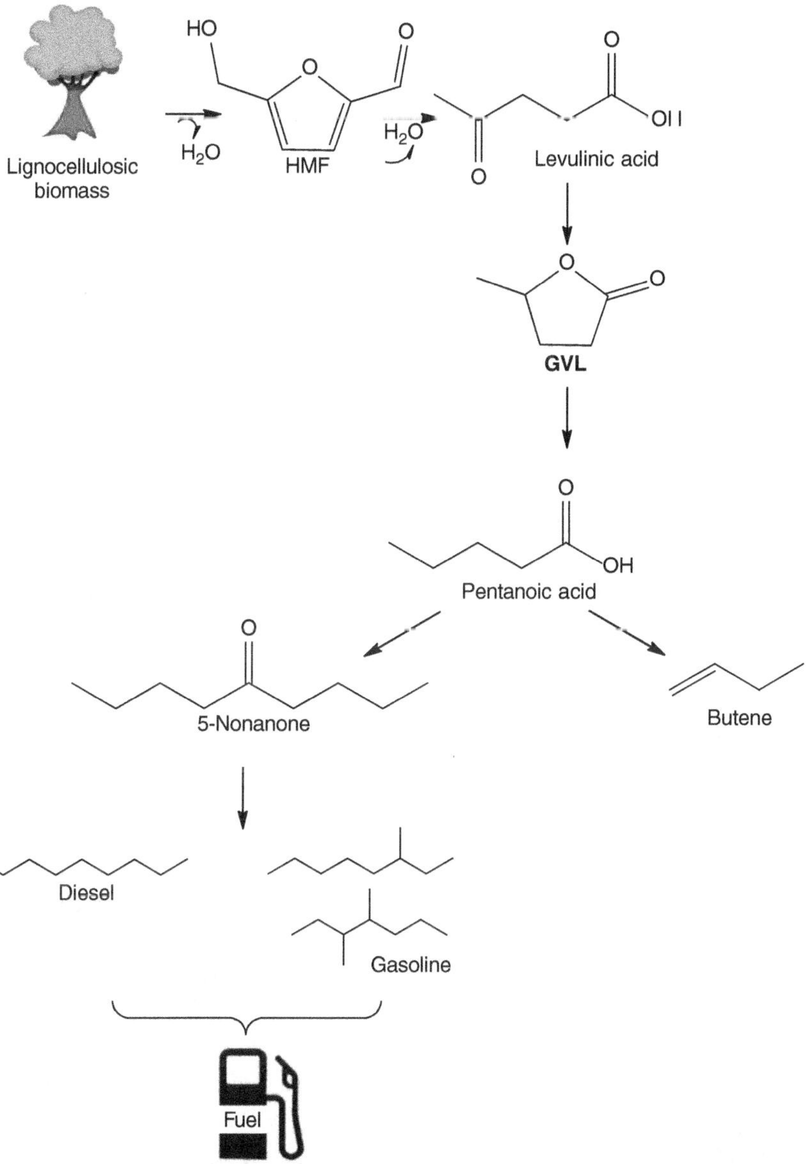

Scheme 4.8 Pathways and applications of the catalytic upgrading of levulinic acid to hydrocarbon fuels and chemicals.

or supercritical CO_2 has been suggested for GVL extraction, although they may be difficult to operate on large scales. Another possibility is the direct processing of the aqueous GVL solutions to produce hydrophobic liquid hydrocarbons that can be used as fuels. Scheme 4.9 presents some pathways for converting GVL into liquid hydrocarbons.

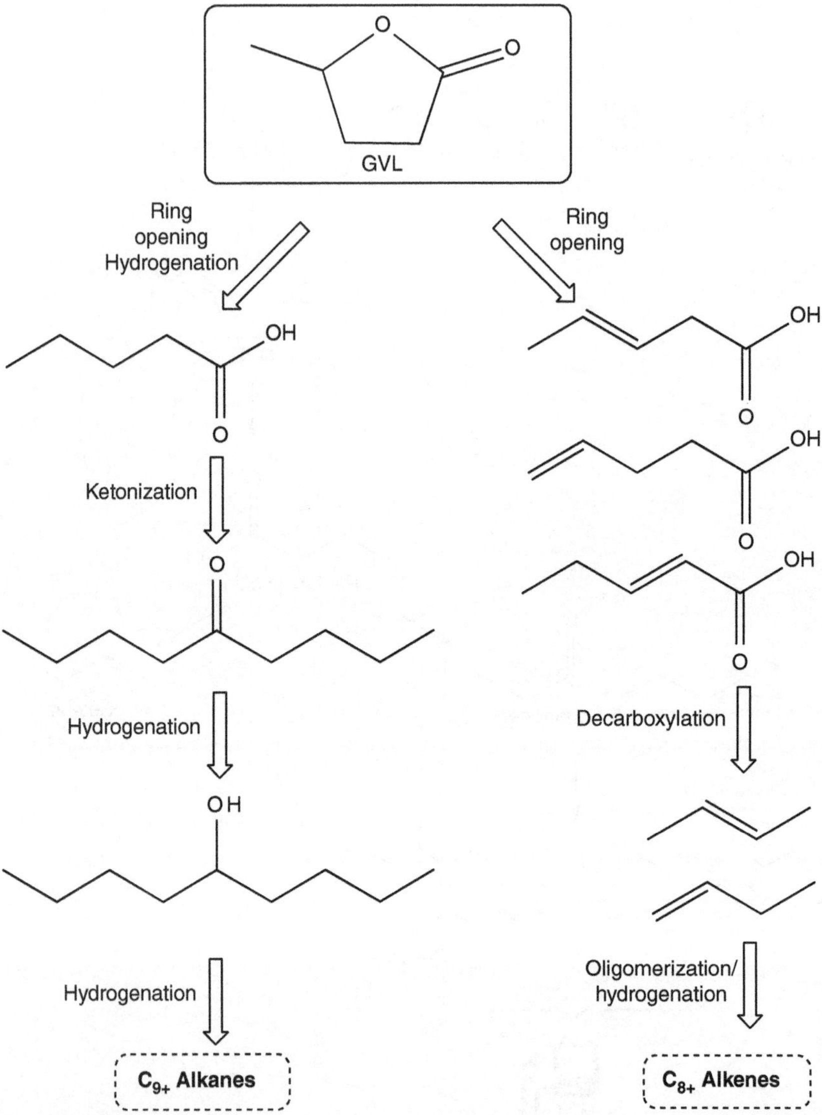

Scheme 4.9 Pathways to convert GVL to liquid fuels. Source: Alonso et al. [47]/Royal Society of Chemistry.

4.4 Homogeneous and Heterogeneous Catalysts for the Efficient Conversion of LA to GVL

Many studies indicated that homogeneous catalysts can give high yields of GVL under relatively mild and controllable reaction conditions [48]. The advantage of the homogeneous catalytic system is the high catalytic performance, but they are difficult to be separated from the medium for reuse. In addition, they are expensive and may present toxicity and poor stability [49].

On the other hand, heterogeneous catalytic systems may be easier separated from the medium for reuse. Therefore, they may present low-processing costs, especially if catalysts based on nonnoble metals are employed [50]. For large-scale processes, heterogeneous catalysts are preferred over homogeneous ones, because of the ease separation and processing of the products [51]. Heterogeneous catalysts could offer great potential for feasible biomass upgrading, attributed to their benefits, such as low-production costs, strong hydrothermal stability, and efficient recovery/reusability. In addition, the use of inexpensive nonprecious metal catalysts may significantly decrease the process costs and pave the way for new technologies of LA hydrogenation [52].

Since the conversion of LA to GVL also involves dehydration, the presence of an acidic catalyst function is necessary. Molecular sieves [26], carbon materials [53], and metal oxides [54] are often used as support to provide acid sites. Thus, bifunctional materials are often employed; whereas the metal particles provide sites for hydrogenation to take place, the acid function of the support may catalyze dehydration of the substrate.

A major problem associated with hydrogenation is the susceptibility of the metal catalyst to poisoning and leaching. The metallic function and the support can react with the carboxylic acid group of LA, causing leaching or dissolution of the metal or support. Thus, the correct choice of the catalytic system is of prime importance to carry out the reaction.

4.4.1 Precious Metal Catalysts

The most efficient catalysts are typically based on precious metal materials [7, 55–57] such as ruthenium (Ru) [58–60], iridium [61–63], and palladium [64]. Among nobles metals, ruthenium seems to give the best results [6, 65].

Ru-based catalysts have been widely applied in the hydrogenation of LA because of the unique ability of the ruthenium atoms to effectively hydrogenate the keto-carbonyl group of LA under mild reaction conditions yielding 4-hydroxyvaleric acid intermediate, which in turn may be dehydrated to GVL [40].

Table 4.3 shows different Ru-supported catalysts for the conversion of LA to GVL in different solvents, such as water [38], alcohols [68, 71], dioxane [20], and

Table 4.3 Catalytic hydrogenation of LA to GVL on Ru-based catalysts.

Catalyst	Solvent	Reaction conditions	Yield	References
Ru/Nb_2O_5	Dioxane	200 °C, 40 bar H_2, 4 h	C = 72%	[6]
			S_{GVL} = 72%	
			S_{MTHF} = 2%	
Ru/TiO_2			C = 2%	
			S_{GVL} = 92%	
			S_{PE} = 2%	
Ru/ZSM-5			C = 6%	
			S_{GVL} = 50%	
			S_{PA} = 30%	
			S_{PE} = 15%	
Ru/H-β			C = 29%	
			S_{GVL} = 60%	
			S_{PA} = 28%	
			S_{PE} = 10%	
Ru-NPs	THF	130 °C, 25 bar, 24 h	C = 10%	[66]
			S_{GVL} = 10%	
	Methanol		C = 100%	
			S_{GVL} = 25%	
	Water		C = 100%	
			S_{GVL} = 99%	
	THF	130 °C, 12 bar, 24 h	C = 98%	
			S_{GVL} = 98%	
	Solvent-free		C = 100%	
			S_{GVL} = 95%	
Ru-NHC	Methanol	130 °C, 12 bar	C = 45%	[67]
			Y = 3%	
	THF		C = 1%	
			Y = 1%	
	Water		C = 99%	
			Y = 96%	
	Dioxane		C = 2%	
			Y = 2%	

(Continued)

Table 4.3 (Continued)

Catalyst	Solvent	Reaction conditions	Yield	References
Ru-C	Methanol	130 °C, 20 bar H_2, 2.6 h	C = 99% S = 61% Y = 81%	[68]
Ru-C	Ethanol		C = 75% S = 85% Y = 84%	
Ru-C	Water		C = 99% S = 86% Y = 87%	
Ru-C	1,4-dioxane		C = 99% S = 969% Y = 98%	
Ru-Al_2O_3	Ethanol		C = 38% S = 32% Y = 86%	
Ru-SiO_2	Ethanol		C = 83% S = 77% Y = 93%	
Ru-C	Water	70 °C, 30 bar H_2, 3 h	C = 48% S = 97%	[29]
		70 °C, 5 bar H_2, 3 h	C = 13% S = 99%	
Ru/NHPC	Water	50 °C, 10 bar H_2, 3 h	C = 99% S = 99%	[69]
Ru/C			C = 50% S = 60%	
Ru/AC			C = 39% S = 84%	
3 Ni-Ti-HZSM5	Dioxane	220 °C, 30 bar H_2, 10 h	C = 99% Y_{GVL} = 94% Y_{PA+PE} = 4%	[70]
3 Ni-Ti-MCM			C = 92% Y_{GVL} = 86%	

(Continued)

Table 4.3 (Continued)

Catalyst	Solvent	Reaction conditions	Yield	References
1 Ru-HZSM5			C = 99%	
			Y_{GVL} = 40%	
			Y_{PA+PE} = 48%	
3 Ru-HZSM5			C = 99%	
			Y_{GVL} = 2%	
			Y_{PA+PE} = 85%	
5 Ru-HZSM5			C = 99%	
			Y_{GVL} = 20%	
			Y_{PA+PE} = 65%	
3-Ru-Zr-MCM			C = 100%	
			Y_{GVL} = 90%	
			Y_{PA+PE} = 3%	
3-Ru-Ti-MCM			C = 100%	
			Y_{GVL} = 92%	
			Y_{PA+PE} = 5%	

C = Conversion, S = Selectivity, Y = Yield, AC = activated carbon, PA = pentanoic acid, PE = pentanoic ester.

supercritical CO_2 [72]. Most studies report conversions above 90% at H_2 pressures in the range of 50–100 bar and temperatures ranging from 100 to 240 °C [54, 73, 74].

The catalytic performance of Ru-supported catalysts is greatly affected by the nature of the support and the choice of solvent [6]. The less-acidic catalysts selectively produced GVL as the main product, whereas zeolite-based catalysts may favor the conversion of LA to PA under mild conditions in dioxane as solvent. The strong acid sites of the support accelerate the conversion of LA to GVL and are essential for the subsequent and most difficult step of ring-opening to yield PA.

Ruthenium nanoparticles (Ru-NPs) have been reported [66] in the hydrogenation of LA. The method is phosphine-free, and good results were achieved under mild reaction conditions. *In situ*-generated ruthenium nanoparticle derivatives of Ru-NHC complexes were active at mild conditions (130 °C and 12 bar, respectively) [67]. The catalytic active ruthenium nanoparticles are formed from all Ru-NHC complexes under H_2 atmosphere in water. In organic solvents, complexes with monodentate NHC ligands decompose to Ru-NPs of low catalytic activity, whereas the complexes with bidentate NHC ligands afford catalysts with moderate hydrogenation activities. The reaction rate is independent of the LA concentration but linearly dependent on the hydrogen pressure within 3–15 bar.

A heterogeneous catalyst made of Ru-NPs supported on hierarchically porous N-doped carbon, Ru/NHPC, was highly active, selective, and stable for the catalytic transformation of LA into GVL. The catalyst could be recycled for 13 consecutive runs without showing significant activity decay [69].

Ru-HZSM-5 catalysts were studied in the synthesis of GVL or valeric biofuels [70]. The catalyst showed good selectivity to the ring-opening of GVL in 1,4-dioxane as solvent, leading to pentanoic esters. Neutral supports, like MCM-41 mesoporous silica, had little effect on hydrogenation, whereas HZSM-5 produced small amounts of PA. With the impregnation of Ru metallic particles on mesoporous supports, the conversion approached 100% and the GVL yield was about 90% with PA being formed in small amounts.

4.4.2 Nonprecious Metal Catalyst

Nonprecious metal catalysts are receiving increased attention for GVL synthesis because of their economic viability and environmental sustainability. The performance of some catalysts based on Cu, Ni, Zr, and Fe, as well as combination of metals in the conversion of LA to GVL will be highlighted next.

4.4.2.1 Copper-Based Catalysts

Table 4.4 shows some selected results of LA hydrogenation on copper-based catalysts.

Hydrogenation of LA over Cu/WO_3-ZrO_2 catalyst gives maximum GVL yields of 99% and 94% in water and ethanol, respectively, at 200 °C, 6 hours, and 50 bar [79]. Metal leaching may be the main problem of LA hydrogenation in aqueous solution because of the good solubility of LA in water, leading to dissociation and formation of strong acidic solutions. Copper-based catalysts supported on SiO_2, TiO_2, ZSM-5, Al_2O_3, or SiO_2–Al_2O_3 were tested in the vapor phase hydrogenation of LA [39]. The conversion was around 48% with 80% selectivity to GVL at 250 °C. Copper-based catalysts supported on ZrO_2, Al_2O_3, Cr_2O_3, or BaO were also tested in LA hydrogenation [75, 81]. After five hours of reaction, the LA conversion was 100% on Cu/ZrO_2 and Cu/Al_2O_3 catalysts in water and methanol as solvents, at 200 °C and 35 bar. The selectivity to GVL in water was 100% on both catalysts, whereas on methanol, the selectivity was 90% on Cu/ZrO_2 and 86% over Cu/Al_2O_3. Magnetic Ni/Cu/Al/Fe catalysts were tested in the hydrogenation of LA to GVL at 142 °C under 20 bar of H_2 for three hours. The yield of GVL was 98% and the magnetic catalyst could be recycled for more than five cycles without showing a significant decrease in the performance [56]. Copper–silica

Table 4.4 Hydrogenation of LA to GVL over copper-based catalysts.

Catalyst	Solvent	Reaction conditions	Yield	References
Cu-ZrO_2	Water	200 °C, 500 psi H_2, 5 h	C = 100% S = 100%	[75]
Cu-Al_2O_3	Water		C = 100% S = 100%	
Cu-ZrO_2	Methanol		C = 100% S = 90%	
Cu-Al_2O_3	Methanol		C = 100% S = 86%	
Cu-BaO	Water		C = 12% S = 100%	
Cu-BaO	Methanol		C = 78% S = 41%	
Cu-Cr_2O_3	Water		C = 9% S = 100%	
Cu-Cr_2O_3	Methanol		C = 72% S = 45%	
Cu-Al	Water	200 °C, 70 bar H_2, 10 h	C = 98.3% Y = 87.3%	[76]
Cu-Cr			C > 99% Y = 90.7%	
Cu-Fe			C > 99% Y = 81.5%	
Cu/Al_2O_3	THF	180 °C, 14 bar H_2, 4 h	C = 100% S = 99%	[77]
Cu/SiO_2	Water	250 °C, formic acid, 10 h, vapor phase hydrogenation	C = 48% S = 80%	[39]
Cu/TiO_2			C = 8% S = 25%	
Cu/Al_2O_3			C = 24% S = 77%	

(Continued)

Table 4.4 (Continued)

Catalyst	Solvent	Reaction conditions	Yield	References
Cu/ZSM-5			C = 38% S = 4%	
Cu/Al_2O_3	Water	250 °C, 65 bar H_2, 6 h	C = 75% S = 66%	[78]
Cu/ZSM-5	Ethanol	200 °C, 50 bar H_2, 6 h	C = 100% Y = 3%	[79]
Cu/SBA-15			C = 100% Y = 20%	
Cu/ZrO_2			C = 100% Y = 4%	
Cu/C			C = 100% Y = 22%	
Cu/Al_2O_3			C = 100% Y = 93%	
Cu-Mo/C	Methanol	200 °C, 100 bar H_2, 2 h	C = 95% S = 45%	[80]
Cu-Mo/C	Dioxane		C = 48% S = 100%	
	Methanol		C = 72% S = 45%	

C = Conversion, S = Selectivity, Y = Yield.

nanocomposites were studied in the hydrogenation of LA to GVL and 2-MTHF under vapor-phase conditions and moderate H_2 pressures. At 265 °C and 1 bar, Cu/SiO_2 showed 100% conversion and 94% selectivity to GVL with β-angelica lactone accounting for the remaining 6%. Upon increasing the pressure to 10 bar, the selectivity to GVL achieved 99.9%, but at more elevated pressures, the selectivity to GVL decreases to 93%, whereas the formation of 1,4-PDO could also be observed [82].

4.4.2.2 Nickel-Based Catalysts

The catalytic performance of some nickel-based catalysts is shown in Table 4.5. Nickel catalysts have been widely used in the conversion of LA to GVL because of their high activity and selectivity.

Table 4.5 Nickel-based catalysts for the hydrogenation of LA to GVL.

Catalyst	Solvent	Reaction conditions	Yield	References
Ni-Mo/C	Dioxane	100 °C, 100 bar H_2, 2 h	C = 95% S = 100%	[80]
		200 °C, 100 bar H_2, 2 h	C = 100% S = 100%	
		60 °C, 100 bar H_2, 2 h	C = 81% S = 100%	
		100 °C, 50 bar H_2, 2 h	C = 78% S = 100%	
	Water	200 °C, 100 bar H_2, 2 h	C = 94% S = 84%	
	Methanol		C = 95% S = 41%	
	Dioxane/water (50/50)		C = 85% S = 95%	
	Dioxane/water (80/20)		C = 97% S = 99%	
	Dioxane/water (20/80)		C = 90% S = 86%	
Ni/Al_2O_3	2-Propanol	200 °C, 2-propanol (H donor), 1 h	C = 99% S = 90%	[37]
Ni/SiO_2			C = 95% S = 40%	
Ni/ZnO			C = 92% S = 35%	
Ni/NiO	Dioxane	120 °C, 20 bar H_2, 24 h	C = 100% S = 100%	[83]
Ni/Al_2O_3	Solvent-free	200 °C, 50 bar H_2, 4 h	C = 92% S = 100%	[84]
Ni/Al_2O_3	Water	250 °C, 65 bar H_2, 2 h	C = 100% S = 92%	[78]

(Continued)

Table 4.5 (Continued)

Catalyst	Solvent	Reaction conditions	Yield	References
Ni/C	Water	160 °C, 40 bar H_2, 3 h	C = 100% S = 94%	[17]
Ni/γAl_2O_3	DMF	150 °C, 10 bar H_2, 6 h	C = 24% Y = 13%	[84]
	Acetic acid		C < 1% Y = 100%	
	Methanol		C = 100% Y = 2%	
	Ethanol		C = 74% Y = 9%	
	Propanol		C = 87% Y = 34%	
	Butanol		C = 100% Y = 19%	
Ni/γAl_2O_3	Water	200 °C, 30 bar H_2, 3 h	C = 98% S = 95%	[85]
	Dioxane	180 °C, 30 bar H_2, 2 h	C = 100% S = 99%	
	Dioxane	180 °C, 30 bar H_2, 1 h	C = 82% S = 98%	
Ni/H-ZSM-5	Solvent-free	250 °C, vapor phase 1 bar	C = 100% Y = 92%	[26]
Ni/C	Dioxane	150 °C, 55 bar H_2	C = 2% Y = 20%	[20]
Raney Ni	Methanol	130 °C, 12 bar H_2	C = 18% Y = 30%	[71]
Ni/MoO_x-C	Solvent-free	140 °C, 8 bar H_2	C = 100% Y = 97%	[86]

C = Conversion, S = Selectivity, Y = Yield.

The nickel catalysts are supported on a variety of materials, such as carbon, alumina, or magnesia [84, 85, 87, 88]. Nickel supported on γ-Al_2O_3 showed 92% LA conversion with 100% selectivity to GVL at 200 °C, 50 bar, and four hours of reaction time, under solventless reaction conditions [84].

Pinto et al. studied bimetallic catalysts supported on activated carbon (Ni–Mo/C, Cu–Mo/C, Zn–Mo/C, Fe–Mo/C) in the hydrogenation of LA to GVL. Catalytic tests, using 1,4-dioxane as solvent, showed that the Ni-Mo/C showed the highest conversion and selectivity for GVL at 200 °C, 100 bar, and two hours of reaction. The catalysts based on Zn–Mo and Fe–Mo were practically inactive under the conditions used, whereas the Cu–Mo/C gave poor conversion. When methanol was used as solvent, the selectivity to GVL drastically decreased due to the formation of methyl levulinate, produced from the acid-catalyzed esterification of LA. In water, 1,4-PDO was also formed as by-product [80]. The behavior in water was associated with hydrogen bonding between LA and the water molecules, which may impair cyclization, favoring hydrogenation of the carboxyl group to yield 1,4-PDO.

Nickel on activated carbon (Ni/C) presented GVL yields higher than 94%, being reused for five consecutive cycles without presenting loss of activity [17]. On the other hand, the hydrogenation of LA to GVL over 5% Ni supported on carbon, with 1,4-dioxane as solvent, at 150 °C, 55 bar, and two hour, afforded just 5% conversion and GVL selectivity of less than 20%, being worse in performance than other metallic catalysts, such as Rh, Pd, Ru, Pt, Re [20].

Bimetallic Ni-MoO_x/C [86] and Ni-Cu/Al_2O_3 [78] catalysts were tested at 250 °C and 50–65 bar of H_2 pressures showing GVL yields of more than 90%.

4.4.2.3 Zirconium-Based Catalysts

The performance of some zirconium-based catalysts is presented in Table 4.6. Zirconium is considered a promising nonnoble metal for the hydrogenation of LA. Several studies report the use of Zr-based catalysts, including ZrO_2, ZrO_2-immobilized on silica and on zeolites, for the hydrogenation of LA to GVL, especially because of the acidic nature of ZrO_2, which favors cyclization [87–89, 97]. More recently, Zr-based metal-organic frameworks (MOFs) have also received great attention in the conversion of LA to GVL [97].

4.4.2.4 Iron-Based Catalysts

Huang et al. reported a novel TiO_2-supported Fe–Re bimetallic catalyst system, which is highly active in the hydrogenation of LA to GVL under mild conditions. Fe/TiO_2 showed GVL yield of less than 1%. Using Fe–Re bimetallic catalysts, it was possible to obtain much higher catalytic performance. With Fe–Re (1 : 1)/TiO_2 catalyst, 12% yield of GVL was obtained at 14% conversion. Catalytic performance

Table 4.6 Zirconium-based catalysts for the hydrogenation of LA to GVL.

Catalyst	Solvent	Reaction conditions	Yield	References
ZrO_2	2-butanol	120 °C, 2-butanol, 11 h	C = 52% S = 22%	[89]
ZrO_2/SBA-15	2-propanol	250 °C, 2-propanol, continuous	C = 99% S = 93%	[90]
	2-butanol	250 °C, 2-butanol, continuous	C = 100% S = 96%	
ZrO_2/β zeolite	2-pentanol	118 °C, 2-pentanol, 10 h	C = 100% S = 96%	[91]
ZrO_2/β zeolite	2-butanol	120 °C, 2- butanol, 11 h	C = 98% S = 99%	[92]
ZrO_2/SBA-15	2-propanol	150 °C, 2-propanol, 2 h	C = 99% Y = 83%	[93]
Zr-CA	2-propanol	130 °C, 2-propanol, 4 h	C = 100% Y = 96%	[94]
Zr-Fe	2-propanol	180 °C, 2-propanol, 2 h	C = 100% Y = 99%	[95]
Zr-Al-β zeolite	Isopropanol	190 °C, 2 h	Y = 90%	[96]

C = Conversion, S = Selectivity, Y = Yield.

increased with increasing Re–Fe atomic ratio. The best catalytic performance was obtained for Fe–Re (1 : 2)/TiO_2, which gave 17% yield of GVL at 18% conversion. These results highlight the synergy between Fe and Re to improve the conversion of LA [98].

Song and coworkers reported the hydrogenation of LA catalyzed by iron complexes with tweezers ligands [99]. GVL was produced in up to 97% yield under mild reaction conditions. When methanol was used as a solvent, the Fe complex with the ligand bis(diisopropyl-phosphinomethyl)pyridine produced GVL in 90% yield within two hours, whereas aliphatic PNP-Fe gave 27% yield of GVL. The Fe complex having 2,6-diaminopyridine as ligand showed 97% yield of GVL. The hydrogenation in ethanol, propanol, THF, and water gives GVL within 90–94% yields. With the use of pure THF, the yield dropped to 15% in 12 hours, indicating the need for a protic solvent.

Table 4.7 shows some examples of LA hydrogenation over iron-based catalysts.

Table 4.7 Iron-based catalysts for the hydrogenation of LA to GVL.

Catalyst	Solvent	Reaction conditions	Yield	References
Fe/TiO_2	Water	140 °C, 40 bar H_2, 4 h	Y < 1%	[98]
			C = 4%	
Fe-Re (1 : 1)/TiO_2			Y = 12%	
			C = 14%	
Fe-Re (1 : 2)/TiO_2			Y = 17%	
			C = 18%	
Fe-1[a,d,e]	Methanol	100 °C, 50 atm H_2, 2 h	Y = 90%	[99]
Fe-2[b,d,e]		100 °C, 50 atm H_2, 5 h	Y = 27%	
Fe-3[c,d,e]		100 °C, 50 atm H_2, 2 h	Y = 97%	
Fe-3[c,d,e]	Ethanol	100 °C, 50 atm H_2, 5 h	Y = 92%	
Fe-3[c,d,e]	Propanol	100 °C, 50 atm H_2, 2 h	Y = 94%	
Fe-3[c,d,e]	Water	100 °C, 50 atm H_2, 2 h	Y = 90%	
Fe-3[c,d,e]	THF	100 °C, 50 atm H_2, 12 h	Y = 15%	
Fe-Mo/C	Dioxane	200 °C, 100 atm H_2, 2 h	C = 0%	[80]
			S = 0%	
Fe-Mo/C	Water		C = 0%	
			S = 0%	
Fe-MoO_x/C	Solvente-free	140 °C, 8 bar H_2, 5 h	C = 0%	[86]
			S = 0%	

a) Fe-1 (bis(diisopropyl-phosphinomethyl)pyridine ligand.
b) Fe-2 (aliphatic PNP-Fe 2).
c) Fe-3 (2,6-diaminopyridine scaffold).
d) [Fe] (0.05 mol%).
e) KOH (5 mmol).

4.5 Heterogeneous Catalysts for the Conversion of LA and GVL to 1,4-PDO and 2-MTHF

Copper catalysts were tested in the hydrogenation of LA to 1,4-PDO [1, 4, 6, 32–35]. Cobalt [100] and nickel [101, 102] catalysts are also used for this purpose. Ni-MoO_x/Al_2O_3 catalysts were used in the conversion of GVL into 1,4-PDO and 2-MTHF [103]. Silica-supported cobalt catalysts were tested in the hydrogenation of LA to produce 2-MTHF and 1,4-PDO as minor by-product [104]. The acid-catalyzed ring-opening of GVL to pentenoic acid, followed by its subsequent hydrogenation to PA, was also studied [5]. The correct adjustment of the acid and

Table 4.8 Hydrogenation of levulinic acid using various supported metal catalysts.

Catalyst	Solvent	Reaction conditions	Yield	References
Pt/MoO_3	Water	130 °C, 50 bar, 6 h	C > 99% Y_{GVL} = 16% $Y_{1,4\text{-}PDO}$ = 74% $Y_{2\text{-}PeOH}$ = 3% $Y_{1\text{-}PeOH}$ = 2% Y_{MTHF} = trace	[105]
Pt/WO_3			C > 99% Y_{GVL} = 88% $Y_{1,4\text{-}PDO}$ = 12% Y_{MTHF} = trace	
Pt/Nb_2O_5			C > 99% Y_{GVL} = 97% $Y_{1,4\text{-}PDO}$ = 3% Y_{MTHF} = trace	
Rh/MoO_3			C > 99% Y_{GVL} = 87% $Y_{1,4\text{-}PDO}$ = 13% Y_{MTHF} = trace	
Pt-Mo/HAP		130 °C, 50 bar, 1 h	C > 99% Y_{GVL} = 67% $Y_{1,4\text{-}PDO}$ = 24% $Y_{2\text{-}PeOH}$ = 2%	
		130 °C, 50 bar, 6 h	C > 99% Y_{GVL} = 23% $Y_{1,4\text{-}PDO}$ = 72% $Y_{2\text{-}PeOH}$ = 3%	
		130 °C, 50 bar, 12 h	C > 99% Y_{GVL} = trace $Y_{1,4\text{-}PDO}$ = 93% $Y_{2\text{-}PeOH}$ = 4%	

(Continued)

Table 4.8 (Continued)

Catalyst	Solvent	Reaction conditions	Yield	References
		130 °C, 30 bar, 24 h	C > 99% Y_{GVL} = trace $Y_{1,4\text{-}PDO}$ = 93% $Y_{2\text{-}PeOH}$ = 5%	
Pt/HAP		130 °C, 50 bar, 12 h	C > 99% Y_{GVL} > 99% $Y_{1,4\text{-}PDO}$ = trace $Y_{2\text{-}PeOH}$ = 5%	
MoO_x/HAP			C > 99% Y_{GVL} = trace $Y_{1,4\text{-}PDO}$ = trace	
Pt-Mo/Al_2O_3			C > 99% Y_{GVL} = 68% $Y_{1,4\text{-}PDO}$ = 32%	
Pd/ZrO_2		200 °C, 60 bar, 24 h	C = 5% S_{GVL} = 0% $S_{1,4\text{-}PDO}$ = 0% S_{MTHF} = 0%	[106]
Pd-Cu/ZrO_2			C = 100% S_{GVL} = 70% $S_{1,4\text{-}PDO}$=25% S_{MTHF} = 5%	
Pd-Cu/Al_2O_3			C = 100% S_{GVL} = 100%	
Cu/ZrO_2			C = 100% S_{GVL} = 70% $S_{1,4\text{-}PDO}$=25% S_{MTHF} = 5%	
Ru/C	Solvent-free	190 °C, 120 bar, 4 h	C = 91% $Y_{2\text{-}BuOH}$ = 28% $Y_{1,4\text{-}PDO}$ = 0%	[42]

(Continued)

Table 4.8 (Continued)

Catalyst	Solvent	Reaction conditions	Yield	References
			$Y_{MTHF} = 31\%$ $Y_{2\text{-PeOH}} = 23\%$	
		190 °C, 100 bar, 24 h	$C > 99\%$ $Y_{2\text{-BuOH}} = 36\%$ $Y_{1,4\text{-PDO}} = 0\%$ $Y_{MTHF} = 43\%$ $Y_{2\text{-PeOH}} = 7\%$	
10 %Cu/ZrO_2-OG	Ethanol	200 °C, 60 bar, 6 h	$C = 38\%$ $S_{1,4\text{-PDO}} = 97\%$ $S_{MTHF} = 3\%$	

metal functions is of prime importance to control the formation of pentenoic acid or 2-MTHF.

The hydrogenation of LA was studied over Pt-supported catalysts at 50 bar and 190 °C in water. The catalyst supported on molybdenum oxide (Pt/MoO_3) produced 1,4-PDO as the major product in 74% yield. The other products were GVL, with 16% yield, 2-pentanol, with 3% yield, and 1-pentanol, with 2% yield after six hours. The yield of 1,4-PDO increased to 83% after 12 hours of reaction [105].

Cu/ZrO_2 and Pd-Cu/ZrO_2 were able to catalyze the hydrogenation of LA to 1,4-PDO and 2-MTHF in water at 200 °C, 60 bar, and 24 hours [106]. Table 4.8 shows some selected results of LA hydrogenation of LA.

4.6 Types of Hydrogenating Agents

Typically, the hydrogenation of LA is carried out over noble metal catalysts with molecular H_2 as the main hydrogenating agent [6, 68, 107]. Nevertheless, these conditions pose some challenges for the large-scale implementation of the process [108, 109]. Besides H_2, the hydrogenation of LA to GVL can be carried out with formic acid (FA) [64] as hydrogenating agent. In addition, an alternative process named catalytic transfer hydrogenation (CTH) can also be used, employing alcohols as hydrogen donors and nonnoble metal catalysts, under relatively mild conditions [87, 110–112].

Figure 4.3 shows a comparison between the direct hydrogenation with H_2 and using FA as hydrogenating agent. In the presence of some metal catalysts, FA decomposes to CO_2 and H_2.

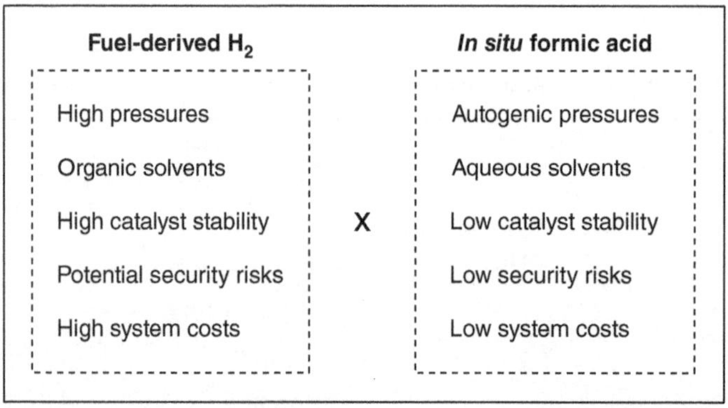

Figure 4.3 Main differences between LA hydrogenation to GVL using H_2 and formic acid.

The CTH process, using alcohols as H-donors, has been proposed for scalable GVL production with economic competitiveness and sustainability, because the reaction occurs under milder conditions without the need for H_2 or noble metal catalysts. Instead, alcohols are used as low-cost and renewable H-donors. Additionally, the alcoholic medium derived from the alcoholysis of sugars can potentially be used as H-donors, thereby leading to energy-saving and cost-effective GVL production.

Secondary alcohols, such as 2-propanol, 2-butanol, and cyclohexanol are the most promising H-donors for CTH, exhibiting 100% conversion of LA and above 90% yield of GVL on ZrO_2/SBA-15 catalyst [90]. The CTH can be considered a Meerwein–Pondorf–Verley (MPV) reduction. For instance, alkyl levulinates can be subjected to MPV reduction to produce GVL on ZrO_2 in the presence of secondary alcohols [89]. This procedure can be an economical alternative to producing GVL from LA, where precious metals and H_2 are avoided and substituted by low-cost catalysts and alcohols. The direct MPV reduction of LA on ZrO_2 is not recommended, as carboxylic acids impair MPV reduction to GVL.

A correlation between the reducing capacity of the alcohols and the activity for MPV reduction has been reported [112]. The order of reactivity is methanol < ethanol < 1-butanol < 2-butanol = 2-propanol, which reinforces the better H-donor ability of secondary alcohols. FA has been employed in the reduction of carbonyl groups [113]. Among the main advantages of the uses of FA are the relatively low toxicity and the good miscibility of LA and water. FA is considered a liquid hydrogen storage system, because it can be decomposed under mild reaction conditions and in the presence of suitable catalysts [114], affording H_2 directly into the medium [115].

The reduction of LA with FA consists of two processes: the selective decomposition of FA to H_2 and the subsequent hydrogenation of LA. The decomposition of

FA is critical to the overall catalytic performance, as it is related to the hydrogen supply. When FA is used as hydrogen donor, two routes can occur: (i) dehydration of FA to produce H_2O and CO, and (ii) dehydrogenation of FA to afford H_2 and CO_2 [116] (Scheme 4.10).

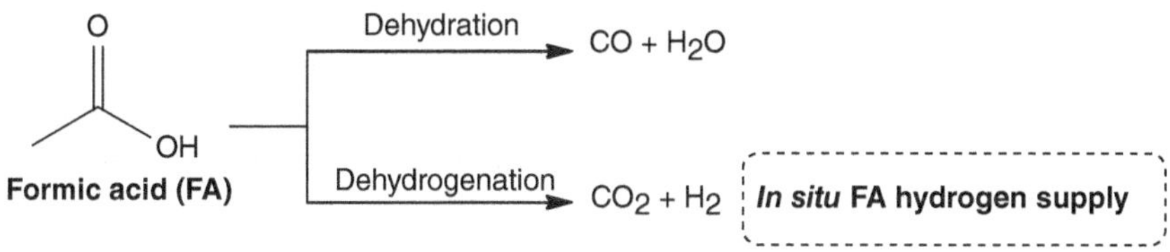

Scheme 4.10 Possible routes of FA decomposition.

The first reaction is not interesting, because it consumes FA without yielding H_2, yet forming CO, which is a well-known metal catalyst poison [117]. Therefore, to increase the yield of GVL, an efficient and selective catalyst system to undergo FA dehydrogenation should be used.

4.7 Patent Search of LA Hydrogenation

The patent search was carried out considering the last 10 years and used "Levulinic acid hydrogenation" as keywords in the title or abstract, yielding 43 documents.

A gas phase process of LA hydrogenation is described in the patent US2786852A [118]. The process occurs upon vaporizing LA at 200 °C, which, together with H_2, is passed over Cu/Cr_2O_3 catalyst at atmospheric pressure. The GVL yield was reported to be 100% under these conditions.

The EP 2537840A1 [11] describes a process for the preparation of GVL from the reaction of LA and hydrogen, using a solid Ru catalyst in the presence of water. The selectivity was higher than 99% with 73% conversion.

The CN1096511304A Patent [119] proposes a method to synthesize GVL through the catalytic hydrogenation of LA over Ag/ZrO_2 catalyst at 220 °C and 40 bar of hydrogen pressure. The LA conversion reached 99.5% and the GVL selectivity can be as high as 99.8%. The choice of Ag/ZrO_2 catalyst was indicated as being cheaper than precious metal catalysts, being also subjected to *in situ* reductions.

The conversion of LA to GVL was reported in the US Patent 6617 464B2 [120]. Different catalysts (Ir, Pd, Pt, Re, Rh, Ru supported on carbon, SiO_2, and Al_2O_3), capable of carrying out the hydrogenation and ring closure were studied. The process was carried out in the liquid phase, using dioxane as solvent, reaching 96%

selectivity to GVL and 100% conversion to LA on 5% Rh/C at 48 bar, 215 °C, and two hours.

The CA2769426A1 Patent [121] reports the hydrogenation of LA to GVL at temperatures ranging from 50 to 350 °C and pressures varying from 2 to 50 bar. GVL is further hydrogenated to VA, or transesterified to afford ethyl pentanoate, in a single vessel operating under reactive distillation conditions.

The US5883266 Patent [122] reported GVL production from LA hydrogenation over bifunctional metal catalysts comprising Cu, Ni, or Re, as well as combinations thereof. The support may be preferably carbon, but alumina and magnesium silicate can also be used.

4.8 Conclusion

Among the various options for LA valorization, catalytic hydrogenation is considered one of the most important and is gaining interest in recent years. Products such as advanced biofuels, as well as fine chemicals, solvents, gasoline, and food additives can be produced from LA hydrogenation processes.

GVL is a promising renewable platform molecule that can be used as a solvent, mainly in processes involving lignocellulosic biomass. GVL also offers new opportunities to replace petroleum-derived feedstock. The hydrogenation of LA can also produce VA, 1,4-PDO, 2-MTHF, 2-pentanol, and 2-butanol, among others, being a multifunctional process of transformation of AL.

References

1 Bozell, J.J. and Petersen, G.R. (2010). Cutting-edge research for a greener sustainable future Technology development for the production of biobased products from biorefinery carbohydrates — The US Department of Energy's "Top 10" revisited. *Green Chem.* 12 (4): 539–554.

2 Pasquale, G., Vázquez, P., Romanelli, G., and Baronetti, G. (2012). Catalytic upgrading of levulinic acid to ethyl levulinate using reusable silica-included Wells-Dawson heteropolyacid as catalyst. *Catal. Commun.* 18: 115–120.

3 Mehdi, H., Fábos, V., Tuba, R. et al. (2008). Integration of homogeneous and heterogeneous catalytic processes for a multi-step conversion of biomass: from sucrose to levulinic acid, γ-valerolactone, 1,4-pentanediol, 2-methyl-tetrahydrofuran, and alkanes. *Top. Catal.* 48 (1–4): 49–54.

4 Geilen, F.M.A., Engendahl, B., Harwardt, A. et al. (2010). Selective and flexible transformation of biomass-derived platform chemicals by a multifunctional catalytic system. *Angew. Chem. Int. Ed.* 49 (32): 5510–5514.

5 Lange, J.P., Price, R., Ayoub, P.M. et al. (2010). Valeric biofuels: a platform of cellulosic transportation fuels. *Angew. Chem. Int. Ed.* 49 (26): 4479–4483.

6 Luo, W., Deka, U., Beale, A.M. et al. (2013). Ruthenium-catalyzed hydrogenation of levulinic acid: influence of the support and solvent on catalyst selectivity and stability. *J. Catal.* 301: 175–186.

7 Alonso, D.M., Wettstein, S.G., and Dumesic, J.A. (2013). Gamma-valerolactone, a sustainable platform molecule derived from lignocellulosic biomass. *Green Chem.* 15 (3): 584–595.

8 Horváth, I.T., Mehdi, H., Fábos, V. et al. (2008). γ-Valerolactone—a sustainable liquid for energy and carbon-based chemicals. *Green Chem.* 10 (2): 238–242.

9 Fegyverneki, D., Orha, L., Láng, G., and Horváth, I.T. (2010). Gamma-valerolactone-based solvents. *Tetrahedron* 66 (5): 1078–1081.

10 Van Der Waal, J.C. and De Jong, E. (2016). Avantium chemicals: the high potential for the levulinic product tree. *Ind. Biorenewables A Pract. Viewp.* 4: 97–120.

11 Castelijns, A.M.C.F., Janssen, M.C.C., Vaessen, H.W.L.M. Process to produce valerolactone from levulinic acid. WO/2012/175439.

12 Gallo, J.M.R., Alonso, D.M., Mellmer, M.A., and Dumesic, J.A. (2013). Production and upgrading of 5-hydroxymethylfurfural using heterogeneous catalysts and biomass-derived solvents. *Green Chem.* 15 (1): 85–90.

13 Bond, J.Q., Martin Alonso, D., West, R.M., and Dumesic, J.A. (2010). γ-Valerolactone ring-opening and decarboxylation over SiO_2/Al_2O_3 in the presence of water. *Langmuir* 26 (21): 16291–16298.

14 Upare, P.P., Lee, J.M., Hwang, Y.K. et al. (2011). Direct hydrocyclization of biomass-derived levulinic acid to 2-methyltetrahydrofuran over nanocomposite copper/silica catalysts. *ChemSusChem* 4 (12): 1749–1752.

15 Wright, W.R.H. and Palkovits, R. (2012). Development of heterogeneous catalysts for the conversion of levulinic acid to γ-valerolactone. *ChemSusChem* 5 (9): 1657–1667.

16 Gao, X., Zhu, S., Dong, M. et al. (2019). Ru nanoparticles deposited on ultrathin TiO_2 nanosheets as highly active catalyst for levulinic acid hydrogenation to Γ-valerolactone. *Appl. Catal. B Environ.* 259 (June): 118076.

17 Fang, S., Cui, Z., Zhu, Y. et al. (2019). In situ synthesis of biomass-derived Ni/C catalyst by self-reduction for the hydrogenation of levulinic acid to Γ-valerolactone. *J. Energy Chem.* 37: 204–214.

18 Van Nguyen, C., Matsagar, B.M., Yeh, J.Y. et al. (2019). MIL-53-NH_2-derived carbon-Al_2O_3 composites supported Ru catalyst for effective hydrogenation of levulinic acid to γ-valerolactone under ambient conditions. *Mol. Catal.* 475 (1): 110478.

19 Ouyang, W., Zhao, D., Wang, Y. et al. (2018). Continuous flow conversion of biomass-derived methyl levulinate into γ-valerolactone using functional metal organic frameworks. *ACS Sustainable Chem. Eng.* 6 (5): 6746–6752.

20 Manzer, L.E. (2004). Catalytic synthesis of α-methylene- γ-valerolactone: a biomass-derived acrylic monomer. *Appl. Catal., A* 272: 249–256.

21 Kamm, M., Gruber, P.R., and Kamm, M. (2012). Biorefineries – industrial processes and products. *Encycl. Ind. Chem.* 5: 659–688.

22 Serrano-ruiz, J.C., Pineda, A., Mariana, A. et al. (2012). Catalytic transformations of biomass-derived acids into advanced biofuels. *Catal. Today* 195: 162–168.

23 Byrne, F., Forier, B., Bossaert, G. et al. (2017). 2,2,5,5-Tetramethyltetrahydrofuran (TMTHF): a non-polar, non-peroxide forming ether replacement for hazardous hydrocarbon solvents. *Green Chem.* 19: 3671–3678.

24 Gürbüz, E.I., Gallo, J.M.R., Alonso, D.M. et al. (2013). Conversion of hemicellulose into furfural using solid acid catalysts in γ-valerolactone. *Angew. Chem* 125: 1270–1274.

25 Cai, C.M., Zhang, T., and Wyman, C.E. (2013). Integrated furfural production as a renewable fuel and chemical platform from lignocellulosic biomass. (April). *J. Chem. Technol. Biotechnol.* 89 (1): 2–10.

26 Mohan, V., Raghavendra, C., Pramod, C.V. et al. (2014). Ni/H-ZSM-5 as a promising catalyst for vapour phase hydrogenation of levulinic acid at atmospheric pressure. *RSC Adv.* 4 (19): 9660–9668.

27 Environ, E., Serrano-ruiz, J.C., and Dumesic, J.A. (2011). Catalytic routes for the conversion of biomass into liquid hydrocarbon transportation fuels. *Energy Environ. Sci.* 4: 83–99.

28 Huang, X., Liu, K., Vrijburg, W.L. et al. (2020). Hydrogenation of levulinic acid to γ-valerolactone over Fe–Re/TiO_2 catalysts. *Appl. Catal. B Environ.* 278 (June): 119314.

29 Galletti, A.M.R., Antonetti, C., De Luise, V., and Martinelli, M. (2012). A sustainable process for the production of γ-valerolactone by hydrogenation of biomass-derived levulinic acid. *Green Chem.* 14 (3): 688–694.

30 Du, X.L., Bi, Q.Y., Liu, Y.M. et al. (2012). Tunable copper-catalyzed chemoselective hydrogenolysis of biomass-derived γ-valerolactone into 1,4-pentanediol or 2-methyltetrahydrofuran. *Green Chem.* 14 (4): 935–939.

31 Sevilla, M. and Fuertes, A.B. (2009). The production of carbon materials by hydrothermal carbonization of cellulose. *Carbon N. Y.* 47 (9): 2281–2289.

32 Yan, K., Yang, Y., Chai, J., and Lu, Y. (2015). Catalytic reactions of γ-valerolactone: a platform to fuels and value-added chemicals. *Appl. Catal., B* 179: 292–304.

33 Simakova, I.L. and Yu, D. (2016). Transformation of bio-derived acids into fuel-like alkanes via ketonic decarboxylation and hydrodeoxygenation: design of multifunctional catalyst, kinetic and mechanistic aspects. *J. Energy Chem.* 25 (2): 208–224.

34 Bereczky, Á., Lukács, K., Farkas, M., and Dóbé, S. (2014). Effect of γ-valerolactone blending on engine performance, combustion characteristics and exhaust emissions in a diesel engine. *Nat. Resour.* 5 (5): 177–191.

35 Fábos, V., Lui, M.Y., Mui, Y.F. et al. (2015). Use of γ-valerolactone as an illuminating liquid and lighter fluid. *ACS Sustainable Chem. Eng.* 3 (9): 1899–1904.

36 Ruppert, A.M., Grams, J., and Je, M. (2015). Titania-supported catalysts for levulinic acid hydrogenation: influence of support and its impact on γ-valerolactone yield. *ChemSusChem* 8 (9): 1–11.

37 Hengne, A.M., Kadu, B.S., Biradar, N.S. et al. (2016). Transfer hydrogenation of biomass-derived levulinic acid to γ-valerolactone over supported Ni catalysts. *RSC Adv.* 6 (64): 59753–59761.

38 Serrano-Ruiz, J.C., Wang, D., and Dumesic, J.A. (2010). Catalytic upgrading of levulinic acid to 5-nonanone. *Green Chem.* 12 (4): 574–557.

39 Lomate, S., Sultana, A., and Fujitani, T. (2018). Vapor phase catalytic transfer hydrogenation (CTH) of levulinic acid to γ-valerolactone over copper supported catalysts using formic acid as hydrogen source. *Catal. Lett.* 148 (1): 348–358.

40 Seretis, A., Diamantopoulou, P., Thanou, I., and Tzevelekidis, P. (2020). Recent advances in ruthenium-catalyzed hydrogenation reactions of renewable biomass-derived levulinic acid in aqueous media. *Front. Chem.* 8 (April): 1–22.

41 Shao, Y., Sun, K., Li, Q. et al. (2019). Copper-based catalysts with tunable acidic and basic sites for the selective conversion of levulinic acid/ester to γ-valerolactone or 1,4-pentanediol. *Green Chem.* 21 (16): 4499–4511.

42 Al-shaal, M.G., Dzierbinski, A., and Palkovits, R. (2014). Solvent-free γ-valerolactone hydrogenation to 2-methyltetrahydrofuran catalysed by Ru/C: a reaction network analysis. *Green Chem.* 16: 1358–1364.

43 Bozell, J.J., Moens, L., Elliott, D.C. et al. (2000). Production of levulinic acid and use as a platform chemical for derived products. *Resour. Conserv. Recy.* 28: 227–239.

44 Bond, J.Q., Alonso, D.M., Wang, D. et al. (2010). Integrated catalytic conversion of γ-valerolactone to liquid alkenes for transportation fuels. *Science* 327: 1110–1114.

45 Christian, R.V. Jr., Brown, H.D., and Hixon, R.M. (1947). Derivatives of γ-valerolactone, 1,4-pentanediol, and 1,4-bis(2-cyanoethoxy)pentane. *J. Am. Chem. Soc.* 69: 1961–1963.

46 Maier, W.F., Roth, W., Thies, I. et al. (1982). Gas phase decarboxylation of carboxylic acids. *Chem. Ber.* 115: 808–812.

47 Alonso, D.M., Bond, J.Q., and Dumesic, J.A. (2010). Catalytic conversion of biomass to biofuels. *Green Chem.* 12: 1493–1513.

48 Chalid, M., Broekhuis, A.A., and Heeres, H.J. (2011). Experimental and kinetic modeling studies on the biphasic hydrogenation of levulinic acid to γ-valerolactone using a homogeneous water-soluble Ru-(TPPTS) catalyst. *J. Mol. Catal. A: Chem.* 341 (1–2): 14–21.

49 Oklu, N.K. and Makhubela, B.C.E. (2018). Highly selective and efficient solvent-free transformation of bio-derived levulinic acid to γ-valerolactone by Ru(II) arene catalyst precursors. *Inorganica Chim. Acta* 482 (Ii): 460–468.

50 Al-Naji, M., Yepez, A., Balu, A.M. et al. (2016). Insights into the selective hydrogenation of levulinic acid to γ-valerolactone using supported mono- and bimetallic catalysts. *J. Mol. Catal. A: Chem.* 417: 145–152.

51 Online, V.A., Ras, E., and Rothenberg, G. (2014). Heterogeneous catalyst discovery using 21st century tools: a tutorial. *RSC Adv.* 4: 5963–5974.

52 Xiao, Z., Zhou, H., Hao, J. et al. (2017). A novel and highly efficient Zr-containing catalyst based on humic acids for the conversion of biomass-derived ethyl levulinate into γ-valerolactone. *Fuel* 193: 322–330.

53 Schuette, H.A. and Thomas, R.W. (1930). Normal valerolactone III. Its preparation by the catalytic reduction of levulinic acid with hydrogen in the presence of platinum oxide. *J. Am. Chem. Soc.* 52 (7): 3010–3012.

54 Tan, J., Cui, J., Deng, T. et al. (2015). Water-promoted hydrogenation of levulinic acid to γ-valerolactone on supported ruthenium catalyst. *ChemCatChem* 7 (3): 508–512.

55 Yan, K., Lafleur, T., Jarvis, C., and Wu, G. (2014). Clean and selective production of γ-valerolactone from biomass-derived levulinic acid catalyzed by recyclable Pd nanoparticle catalyst. *J. Cleaner Prod.* 72: 230–232.

56 Zhang, J., Chen, J., Guo, Y., and Chen, L. (2015). Effective upgrade of levulinic acid into γ-valerolactone over an inexpensive and magnetic catalyst derived from hydrotalcite precursor. *ACS Sustainable Chem. Eng.* 3 (8): 1708–1714.

57 Gonçalves, M., Rodrigues, R., Galhardo, T.S., and Carvalho, W.A. (2016). Highly selective acetalization of glycerol with acetone to solketal over acidic carbon-based catalysts from biodiesel waste. *Fuel* 181: 46–54.

58 Engineering, E.P., Avenue, T.C., and Kong, H. (2014). Selective conversion of levulinic and formic acids to γ-valerolactone with the shvo catalyst. *Organometallics* 33 (1): 181–187.

59 Tukacs, J.M., Novák, M., Dibó, G., and Mika, L.T. (2014). An improved catalytic system for the reduction of levulinic acid to γ-valerolactone. *Catal. Sci. Technol.* 4 (9): 2908–2912.

60 Chowdhury, A.D., Jackstell, R., and Beller, M. (2014). Towards the efficient development of homogeneous catalytic transformation to γ-valerolactone from biomass-derived platform chemicals. *ChemCatChem* 6: 3360–3365.

61 Suriyaprapadilok, N. and Kitiyanan, B. (2011). Synthesis of solketal from glycerol and its reaction with benzyl alcohol. *Energy Procedia* 9: 63–69.

62 Deng, J., Wang, Y., Pan, T. et al. (2013). Conversion of carbohydrate biomass to γ-valerolactone by using water-soluble and reusable iridium complexes in acidic aqueous media. *ChemSusChem* 6 (7): 1163–1167.

63 Links, D.A., Li, W., Xie, J. et al. (2012). Highly ef fi cient hydrogenation of biomass-derived levulinic acid to γ -valerolactone catalyzed by iridium pincer complexes. *Green Chem.* 14 (9): 2388–2390.

64 Ortiz-Cervantes, C., Flores-Alamo, M., and García, J.J. (2015). Hydrogenation of biomass-derived levulinic acid into γ-valerolactone catalyzed by palladium complexes. *ACS Catal.* 5 (3): 1424–1431.

65 Cao, S., Monnier, J.R., Williams, C.T. et al. (2015). Rational nanoparticle synthesis to determine the effects of size, support, and K dopant on Ru activity for levulinic acid hydrogenation to γ-valerolactone. *J. Catal.* 326: 69–81.

66 Ortiz-cervantes, C. and García, J.J. (2013). Hydrogenation of levulinic acid to γ-valerolactone using ruthenium nanoparticles. *Inorganica Chim. Acta* 397: 124–128.

67 Tay, B.Y., Wang, C., Phua, P.H. et al. (2016). Selective hydrogenation of levulinic acid to γ-valerolactone using in situ generated ruthenium nanoparticles derived from Ru-NHC complexes. *Dalton Trans.* 45 (8): 3558–3563.

68 Al-Shaal, M.G., Wright, W.R.H., and Palkovits, R. (2012). Exploring the ruthenium catalysed synthesis of γ-valerolactone in alcohols and utilisation of mild solvent-free reaction conditions. *Green Chem.* 14 (5): 1260–1263.

69 Wei, Z., Li, X., Deng, J. et al. (2018). Improved catalytic activity and stability for hydrogenation of levulinic acid by Ru/N-doped hierarchically porous carbon. *Mol. Catal.* 448 (February): 100–107.

70 Yi, Z., Hu, D., Xu, H. et al. (2020). Metal regulating the highly selective synthesis of gamma-valerolactone and valeric biofuels from biomass-derived levulinic acid. *Fuel* 259 (September 2019): 3–6.

71 Yan, Z.P., Lin, L., and Liu, S. (2009). Synthesis of γ-valerolactone by hydrogenation of biomass-derivedLevulinic acid over Ru/C catalyst. *Energy Fuels* 23 (8): 3853–3858.

72 Bourne, R.A., Stevens, J.G., Ke, J., and Poliakoff, M. (2007). Maximising opportunities in supercritical chemistry: the continuous conversion of levulinic acid to γ-valerolactone in CO_2. *Chem. Commun.* 2: 4632–4634.

73 Yan, Z., Lin, L., Liu, S. et al. (2009). Synthesis of γ-valerolactone by hydrogenation of biomass-derived levulinic acid over Ru/C catalyst. *Energy Fuels* 48 (5): 3853–3858.

74 Chalid, M., Heeres, H.J., and Broekhuis, A.A. (2012). Green polymer precursors from biomass-based levulinic acid. *Procedia Chem.* 4: 260–267.

75 Hengne, A.M. and Rode, C.V. (2012). Cu–ZrO_2 nanocomposite catalyst for selective hydrogenation of levulinic acid and its ester to γ-valerolactone. *Green Chem.* 14 (4): 1064–1072.

76 Yan, K., Liao, J., Wu, X., and Xie, X. (2013). A noble-metal free Cu-catalyst derived from hydrotalcite for highly efficient hydrogenation of biomass-derived furfural and levulinic acid. *RSC Adv.* 3 (12): 3853–3856.

77 Zhang, L., Mao, J., Li, S. et al. (2018). Hydrogenation of levulinic acid into gamma-valerolactone over in situ reduced CuAg bimetallic catalyst: strategy and mechanism of preventing Cu leaching. *Appl. Catal., B* 232 (February): 1–10.

78 Obregón, I., Corro, E., Izquierdo, U. et al. (2014). Levulinic acid hydrogenolysis on Al_2O_3-based Ni-Cu bimetallic catalysts. *Chin. J. Catal.* 35 (5): 656–662.

79 Xu, Q., Li, X., Pan, T. et al. (2016). Supported copper catalysts for highly efficient hydrogenation of biomass-derived levulinic acid and γ-valerolactone. *Green Chem.* 18 (5): 1287–1294.

80 Pinto, B.P., Luísa, A., Christiano, L.F., and Mota, C.J.A. (2017). Hydrogenation of levulinic acid (LA) to γ-valerolactone (GVL) over Ni–Mo/C catalysts and water-soluble solvent systems. *Catal. Lett.* 147: 751–757.

81 Yuan, J., Li, S., Yu, L. et al. (2013). Copper-based catalysts for the efficient conversion of carbohydrate biomass into g-valerolactone in the absence of externally added hydrogen. *Energy Environ. Sci.* 6: 3308.

82 Upare, P.P., Hwang, K., Lee, J., and Hwang, W. (2015). Chemical conversions of biomass-derived platform chemicals over copper–silica nanocomposite catalysts. (Table 1). *ChemSusChem* 4: 2345–2357.

83 Song, S., Yao, S., Cao, J. et al. (2017). Heterostructured Ni/NiO composite as a robust catalyst for the hydrogenation of levulinic acid to γ-valerolactone. *Appl. Catal., B* 217: 115–124.

84 Hengst, K., Schubert, M., Carvalho, H.W.P. et al. (2015). Synthesis of γ-valerolactone by hydrogenation of levulinic acid over supported nickel catalysts. *Appl. Catal., A* 502 (1): 18–26.

85 Fu, J., Sheng, D., and Lu, X. (2016). Hydrogenation of levulinic acid over nickel catalysts supported on aluminum oxide to prepare γ-valerolactone. *Catalysts* 6 (6).

86 Shimizu, K., Kanno, S., and Kon, K. (2014). Hydrogenation of levulinic acid to γ-valerolactone by Ni and MoO_x co-loaded carbon catalysts. *Green Chem.* 16: 3899–3903.

87 Mohan, V., Venkateshwarlu, V., and Pramod, C.V. (2014). Vapour phase hydrocyclisation of levulinic acid to γ-valerolactone over supported Ni catalysts. *Catal. Sci. Technol.* 4: 1253–1259.

88 Lv, J., Rong, Z., Wang, Y. et al. (2015). Highly efficient conversion of biomass-derived levulinic acid into γ-valerolactone over Ni/MgO catalyst. *RSC Adv.* 5 (88): 72037–72045.

89 Chia, M. and Dumesic, J.A. (2011). Liquid-phase catalytic transfer hydrogenation and cyclization of levulinic acid and its esters to γ-valerolactone over metal oxide catalysts. *Chem. Commun.* 47 (44): 12233–12235.

90 Enumula, S.S., Gurram, V.R.B., Kondeboina, M. et al. (2016). ZrO_2/SBA-15 as an efficient catalyst for the production of γ-valerolactone from biomass-derived levulinic acid in the vapour phase at atmospheric pressure. *RSC Adv.* 6 (24): 20230–20239.

91 Wang, J., Jaenicke, S., and Chuah, G.K. (2014). Zirconium-Beta zeolite as a robust catalyst for the transformation of levulinic acid to γ-valerolactone via Meerwein-Ponndorf-Verley reduction. *RSC Adv.* 4 (26): 13481–13489.

92 Bui, L., Luo, H., Gunther, W.R., and Román-Leshkov, Y. (2013). Domino reaction catalyzed by zeolites with Brønsted and Lewis acid sites for the production of γ-valerolactone from furfural. *Angew. Chem.* 125 (31): 8180–8183.

93 Kuwahara, Y. and Kaburag, W. (2014). Catalytic conversion of levulinic acid and its esters to γ-valerolactone over silica-supported zirconia catalysts. *Bull. Chem. Soc. Jpn.* 87: 1252–1254.

94 Xue, Z., Jiang, J., Li, G. et al. (2016). Zirconium-cyanuric acid coordination polymer: Highly efficient catalyst for conversion of levulinic acid to γ-valerolactone. *Catal. Sci. Technol.* 6: 5374–5379.

95 Gu, J., Zhang, J., Wang, Y. et al. (2020). Efficient transfer hydrogenation of biomass derived furfural and levulinic acid via magnetic zirconium nanoparticles: experimental and kinetic study. *Ind. Crops Prod.* 145 (June 2019): 112133.

96 Morales, G., Melero, J.A., Iglesias, J. et al. (2019). From levulinic acid biorefineries to γ-valerolactone (GVL) using a bi-functional Zr-Al-Beta catalyst. *Reac. Chem.Eng.* 4: 1834–1843.

97 Lin, Z., Cai, X., Fu, Y. et al. (2017). Cascade catalytic hydrogenation – cyclization of methyl levulinate to form γ-valerolactone over Ru nanoparticles supported on a sulfonic acid. *RSC Adv.* 7: 44082–44088.

98 Huang, X., Liu, K., Vrijburg, W.L. et al. (2020). Hydrogenation of levulinic acid to γ-valerolactone over Fe–Re/TiO_2 catalysts. *Appl. Catal. B Environ.* 278 (July): 119314.

99 Yi, Y., Liu, H., Xiao, L.P. et al. (2018). Highly efficient hydrogenation of levulinic acid into γ-valerolactone using an iron pincer complex. *ChemSusChem* 11 (9): 1474–1478.

100 Liu, Z., Yang, Z., Wang, P. et al. (2019). Co-catalyzed hydrogenation of levulinic acid to γ-valerolactone under atmospheric pressure. *ACS Sustain. Chem. Eng.* 7 (22): 18236–18241.

101 Liu, D., Zhang, L., Han, W. et al. (2019). One-step fabrication of Ni-embedded hierarchically-porous carbon microspheres for levulinic acid hydrogenation. *Chem. Eng. J.* 369 (March): 386–393.

102 Kumar, V.V., Naresh, G., Deepa, S. et al. (2016). Influence of W on the reduction behaviour and Brønsted acidity of Ni/TiO_2 catalyst in the hydrogenation of levulinic acid to valeric acid: pyridine adsorbed DRIFTS study. *Applied Catal. A* 531: 169–176.

103 Zhang, G., Li, W., Fan, G. et al. (2019). Controlling product selectivity by surface defects over MoO_x-decorated Ni-based nanocatalysts for γ-valerolactone hydrogenolysis. *J. Catal.* 379: 100–111.

104 Novodárszki, G., Solt, H.E., Valyon, J. et al. (2019). Selective hydroconversion of levulinic acid to γ-valerolactone or 2-methyltetrahydrofuran over silica-supported cobalt catalysts. *Catal. Sci. Technol.* 9: 2291–2304.

105 Mizugaki, T., Nagatsu, T., Togo, K. et al. (2015). Selective hydrogenation of levulinic acid to 1,4-pentanediol in water using a hydroxyapatite-supported Pt–Mo bimetallic catalyst. *ACS Sustain. Chem. Eng.* 3 (11): 2619–2630.

106 Patankar, S.C., Yadav, G.D., and Accepted, J. (2015). Cascade engineered synthesis of γ-valerolactone, 1,4-pentanediol and 2-methyltetrahydrofuran from levulinic acid using novel Pd-Cu/ZrO_2 catalyst in water as solvent. *Green Chem.* 17: 5136–5139.

107 Ding, D., Wang, J., Xi, J. et al. (2014). High-yield production of levulinic acid from cellulose and its upgrading to γ-valerolactone. *Green Chem.* 16: 3846–3853.

108 Sun, M., Xia, J., Wang, H. et al. (2018). An efficient NixZryO catalyst for hydrogenation of bio-derived methyl levulinate to Γ-valerolactone in water under low hydrogen pressure. *Appl. Catal., B* 227 (September 2017): 488–498.

109 Sun, D., Ohkubo, A., Asami, K. et al. (2017). Vapor-phase hydrogenation of levulinic acid and methyl levulinate to Γ-valerolactone over non-noble metal-based catalysts. *Mol. Catal.* 437: 105–113.

110 Kuwahara, Y., Kaburagi, W., Osada, Y., and Fujitani, T. (2017). Catalytic transfer hydrogenation of biomass-derived levulinic acid and its esters to γ-valerolactone over ZrO_2 catalyst supported on SBA-15 silica. *Catal. Today* 281: 418–428.

111 Kumaravel, S., Thiripuranthagan, S., Durai, M. et al. (2020). Catalytic transfer hydrogenation of biomass-derived levulinic acid to γ-valerolactone over Sn/Al-SBA-15 catalysts. *New J. Chem.* 44 (20): 8209–8222.

112 Tang, X., Zeng, X., Li, Z. et al. (2015). In situ generated catalyst system to convert biomass-derived levulinic acid to γ-valerolactone. *ChemCatChem* 7 (8): 1372–1379.

113 Kei, F., Cheung, K., Clarke, A.J. et al. (2010). Kinetic and structural studies on 'tethered' Ru (II) arene ketone reduction catalysts. *Dalton Trans.* 39: 1395–1402.

114 Enthaler, S. (2008). Carbon dioxide—the hydrogen-storage material of the future? *ChemSusChem* 1: 801–804.

115 Boddien, A., Mellmann, D., Gärtner, F. et al. (2011). Efficient dehydrogenation of formic acid using an iron catalyst. *Science* 333: 1733.

116 Ojeda, M. and Iglesia, E. (2009). Formic acid dehydrogenation on Au-based catalysts at near-ambient temperatures. *Angew. Chem.* 121: 4894–4897.

117 Ruppert, A.M., Jędrzejczyk, M., Sneka-Płatek, O. et al. (2016). Ru catalysts for levulinic acid hydrogenation with formic acid as a hydrogen source. *Green Chem.* 18 (7): 2014–2028.

118 Dunlop, A.P., Riverside, Madden, J.W. (1957). Process of preparing gamma valerolactone. US Patent 2,786,852.

119 Jinxia, Z., Xufeng, B., Jingbo, M., and Shenmin, L. (2019) Method for preparing gamma-valerolactone by adopting levulinic acid and utilizing catalytic hydrogenation. CN1096511304A.

120 Manzer, L.E. (2003) Production of 5-methylbutyrolactone from levulinic acid. US 6,617,464 B2.

121 Haan, R.J. and Lange, J.P. (2011) Process for preparing a hydroxyacid or hydroxyester. WO 2011/015645 A2.

122 Elliott, D.C. and Frye, J.G. (1999). Hydrogenated 5-carbon compound and method of making. USOO5883266A patent.

5

Carbonyl Reactions of Levulinic Acid – Ketals and Other Derivatives Formed Upon Reaction with the Carbonyl Group of Levulinic Acid. Production Routes, Technologies, and Main Uses

Levulinic acid is an important intermediate in organic synthesis. In addition to being an easily accessible compound, the levulinic acid molecule has a great synthetic potential due to the presence of two highly reactive electrophilic centers. In addition to the carboxyl acid function, it also has a ketone carbonyl functionality, which can be converted into a variety of derivatives of wide industrial applications, as shown in Scheme 5.1. A more detailed description of some derivatives formed

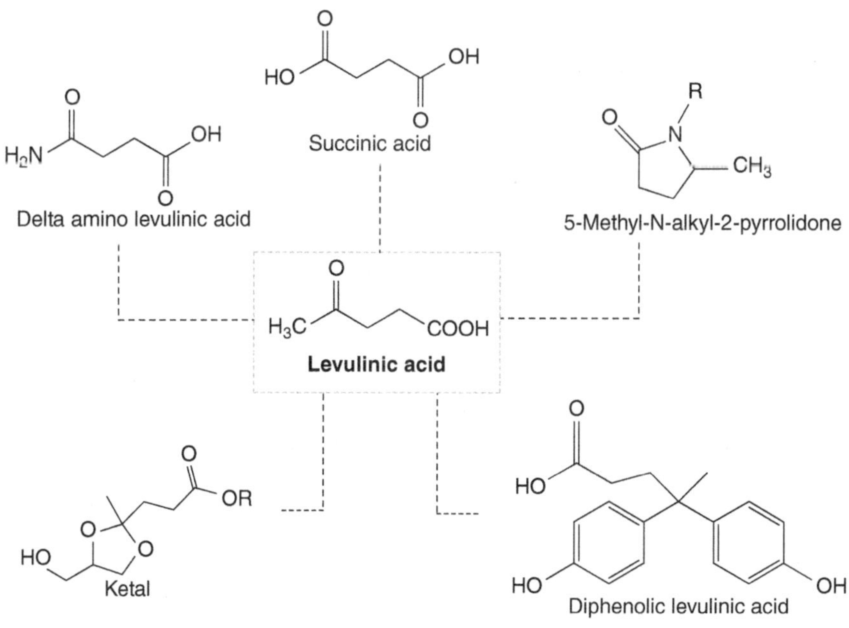

Scheme 5.1 Main derivatives formed upon reaction with the carbonyl group of levulinic acid.

Levulinic Acid: A Sustainable Platform Chemical for Value-Added Products, First Edition.
Claudio J.A. Mota, Ana Lúcia de Lima, Daniella R. Fernandes, and Bianca P. Pinto.

upon reaction with the carbonyl group of levulinic acid can be found throughout the text [1, 2].

5.1 Levulinc Acid Ester Ketals Main Routes

The levulinic acid ester ketals (LEKs) can be obtained from the levulinate esters, or even directly from levulinic acid, in the presence of polyalcohols (or polyols) and homogeneous or heterogeneous acid catalysts [3–5]. The traditional reaction for the formation of LEKs is ketalization, a classic organic chemistry transformation widely used as synthetic strategy, because ketals and acetals are excellent protecting groups. Although well established, there are still few studies focused on the formation of LEKs in the literature [6–8]. The selective ketalization of levulinic acid or its esters is a major challenge because secondary esterification and transesterification products can be formed [3, 9]. In this context, the synthesis of selective catalysts for the formation of LEKs via ketalization is an excellent path for the development of new bio-based building blocks.

The reaction of alcohols with carbonyl compounds forms a ketal, when a ketone is involved, or an acetal when the reaction occurs with an aldehyde. On the other hand, alcohols react with carboxylic acids to afford esters, which may undergo transesterification when reacted with another alcohol [3]. All these reactions require an acid catalyst, especially ketalization/acetalization. The acid catalyst protonates the carbonyl group, making it more reactive toward nucleophilic attack by the alcohol. Ketalization/acetalization is reversible; upon excess of water, the reaction is shifted to the opposite direction, and the ketal/acetal is transformed back into the ketone/aldehyde as well as the respective alcohol.

The most commonly used catalysts in ketalization/acetalization are sulfuric, hydrochloric, and p-toluenesulfonic acids, despite the disadvantages associated with the homogeneous reaction [3]. Ethyl levulinate glycerol ketal, for example, can be obtained with high yield and selectivity at 80–110 °C and 30 mmHg, using sulfuric acid as catalyst [8]. Commercial acid resins have shown satisfactory results in the condensation of ketones with alcohols, with the advantage of being a heterogeneous catalyst system. However, the type and concentration of catalyst used in the ketalization are the same for esterification and transesterification, which may result in poor selectivity and formation of by-products. Zeolites and carbon materials have been reported in the levulinic acid ketalization with polyols. However, the development of selective and sustainable heterogeneous acid catalysts is still a major challenge, which may directly contribute to the valorization of LA for the production of chemicals of high economic value for application in several industrial sectors.

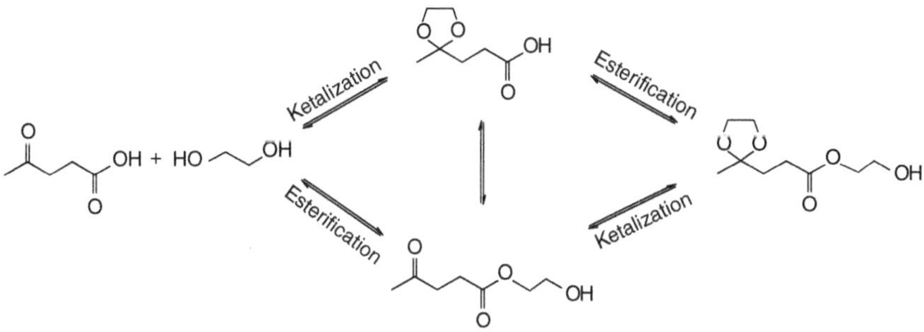

Scheme 5.2 Reaction pathways for the formation of levulinic acid ester ketals from ethylene glycol and levulinic acid.

In the presence of a diol, levulinic acid can lead to two competitive pathways (Scheme 5.2). The first is ketalization, upon reaction with the carbonyl group, whereas the second is a monohydroxylated keto-ester, which originated from the esterification of the carboxyl group. Both products are interconvertible and can yield the cyclic LEK.

Depending on the polyol used, the complexity of the system increases. Therefore, adjustments to the reaction parameters are necessary to improve the selectivity. Glycerol, for example, has two primary hydroxyls and one secondary hydroxyl group, enabling two nucleophilic centers capable of interacting with the carbonyls of the keto-acid, yielding two possible ketalization and esterification products. In addition to these transformations, transesterification between the ester group and the ketal is also possible, since there is a free hydroxyl in the molecule (Scheme 5.3).

The study by Mullen et al. demonstrated the selective formation of cyclic ketals from the reaction of polyalcohols with keto-esters in the presence of acid catalysts. The authors reinforced the possibility of transesterification between the polyalcohol and the keto-ester, as well as the reactivity of the monohydroxylated keto-ester formed, which contains a residual hydroxyl capable of undergoing acyclic ketalization or even an additional transesterification. When ketalization takes place in the presence of polyalcohols containing more than two hydroxyl groups, such as glycerol, the possible number of by-products increases accordingly. Glycerol levulinate ketal (GLKs) contain free hydroxyls, which could undergo additional transesterification with the keto-ester or even intramolecular transesterification. Despite all the possibilities, authors found high selectivity to ketalization products when using lower concentrations of the acid catalysts [3].

Long-chain (or higher) polyols can also be used to form cyclic ketals when, at least, two hydroxyl groups of the polyol are in the 1,2 or 1,3 configurations. Freitas et al. demonstrated this assertion when studying the ketalization of ethyl levulinate with different dodecanediol isomers. Amberlyst-70, H-ZSM-5 zeolite,

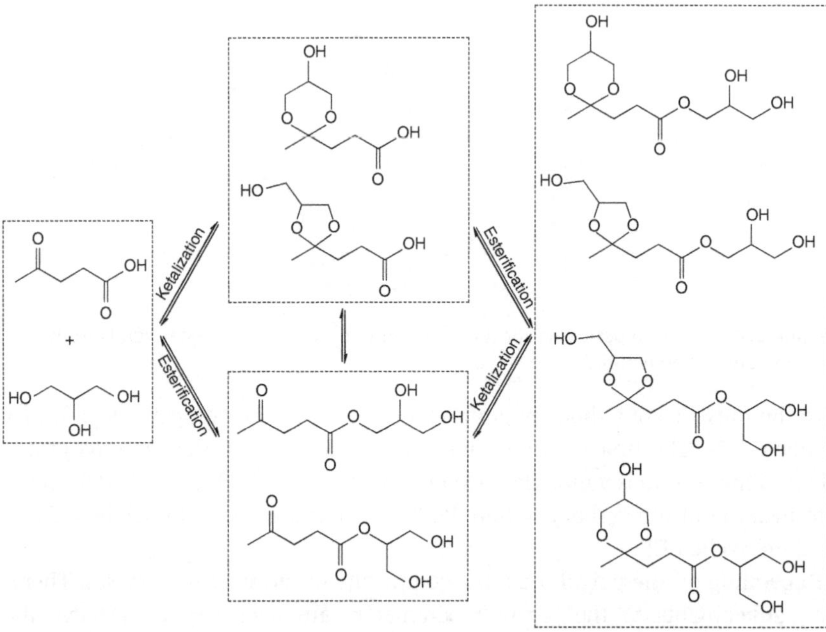

Scheme 5.3 Possible levulinic acid ester ketals formed from glycerol and levulinic acid.

and niobium phosphate were used as heterogeneous acids resulting in good conversion and selectivity to the desired cyclic ketal. Authors also showed that optimization of the reaction parameters, such as catalyst loading, molar ratio of ethyl levulinate to alcohol, and reaction time was fundamental to improving the selectivity. Niobium phosphate, in particular, is a promising system, due to its complete recyclability and good performance even at low catalyst loadings. The synthesized long-chain ketal esters were tested as green surfactants and presented values of surface tension comparable to commercial products [4].

Amberlyst-15 and p-toluene sulfonic acid were used in the condensation of levulinic acid with 1,2-ethanediol (ethylene glycol), 1,2-propanediol (propylene glycol), and 1,3-propanediol, aiming to understand the competition between ketalization and esterification on those systems. When Amberlyst-15 was used as heterogeneous catalyst, ketals and ketal esters were the major products. However, when p-toluene sulfonic acid was employed as homogeneous catalyst, esters and the ketal esters were formed in higher yields. For instance, LA monoester is formed in 75 and 95% yields in reactions with 1,2-propanediol and 1,3-propanediol, respectively [10].

Zeolite Beta was evaluated in the reaction of levulinic acid with different polyalcohols (1,2-propanediol, 1,2-ethanediol, and 1,2,3-propanetriol), using conventional and MW-irradiated reactors. The purpose of the study was to evaluate the selectivity of the intermediates (ester and ketal) and the final product (keto-ester), in addition to reducing the amount of oligomerized or polycondensated products [5].

Konwar et al. studied macro/mesoporous solid carbon materials, derived from Na-lignosulphonate, with highly functionalized surfaces and large amounts of acidic sites, such as –OH, –COOH, and $-SO_3H$ groups. The materials were active for the ketalization/acetalization of glycerol with several bio-based aldehydes and ketones, including methyl levulinate [11].

The reaction between levulinic acid and epoxidized long-chain methyl esters forms ketals with applications as lubricating fluids, surfactants, and fuel additives [12]. The reaction of levulinic acid with methyl 9,10-epoxy-oleate (EMO) may follow two different pathways. In the presence of an acid catalyst, the correspondent LA ketal is formed as the major product, whereas in the absence of a catalyst the opening of the epoxide ring, with the formation of a levulinate, is the preferred pathway (Scheme 5.4) [9].

The reversible reaction of 1,2-O-isopropylidene-glycerol ester (IPGE) with ethyl levulinate, in the presence of tungstophosphoric acid supported on γ-alumina, was evaluated as a strategy to block the free hydroxyl, improving its properties for various uses. The reaction is a classic adaptation of transketalization, where a particular ketal reacts with an alcohol to form a different ketal and a new alcohol. In this particular case, IPGE reacts with the carbonyl moiety of ethyl levulinate to afford a levulinate ketal and acetone, avoiding transesterification side reactions (Scheme 5.5). Acetone is continuously withdrawn from the reaction mixture to shift equilibrium [13].

5.1.1 Levulinic Acid Ester Ketals Main Uses

LEK can present hydrophilic and hydrophobic chemical groups. Depending on the type of ketal, these characteristics together with their specific properties allow great potential for application in several sectors. GLK synthesized from glycerol and ethyl levulinate, for example, has applications in polymers. Some of its properties include freezing point below −60 °C, boiling point of 286 °C (at 1 atm.), viscosity of 37–40 cP at 25 °C, and thermal stability up to 300 °C [14]. The same type of ketal also finds applications in different types of formulations, due to its solubilizing power [15]. Figure 5.1 shows some applications of LEKs.

The presence of hydrophilic groups, such as hydroxyls and hydrophobic moieties, such as hydrocarbon chains associated with the ester group, in a single structure makes LEKs miscible with water and organic compounds [15, 16], being a potential green solvent or cosolvent. As surfactants, the main potential application is in cleaning formulations, lotions, and paints [15]. Some of the sectors that

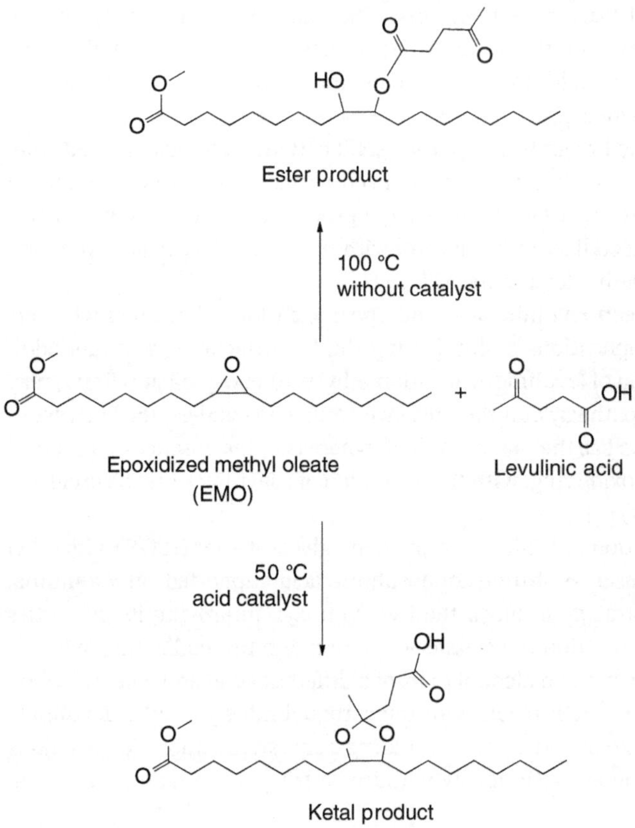

Scheme 5.4 Synthesis of branched oleochemicals from epoxidized methyl oleate (EMO) and levulinic acid [9].

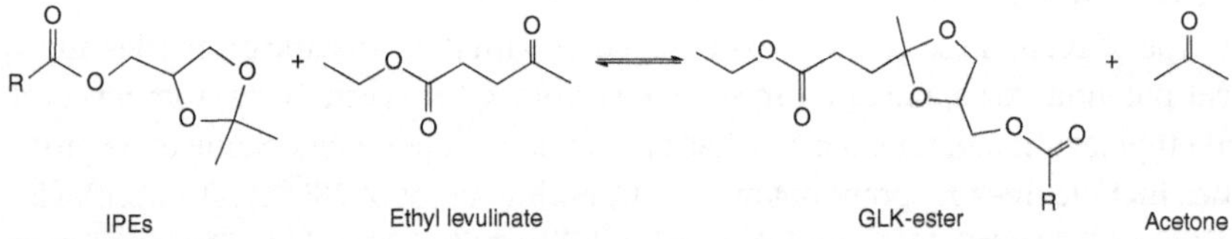

Scheme 5.5 Formation of glycerol levulinate ketal ester (GLK-ester) from the reaction of ethyl levulinate with IPEs, where R corresponds to propyl and hexyl groups.

may use LEKs include pharmaceuticals, cosmetics, and personal care. Different types of LEKs, including GLKs, are included in formulations with the application as medical active agent, solid lip gloss, skin whitening, antiperspirant stick, and hair dyes [15, 17].

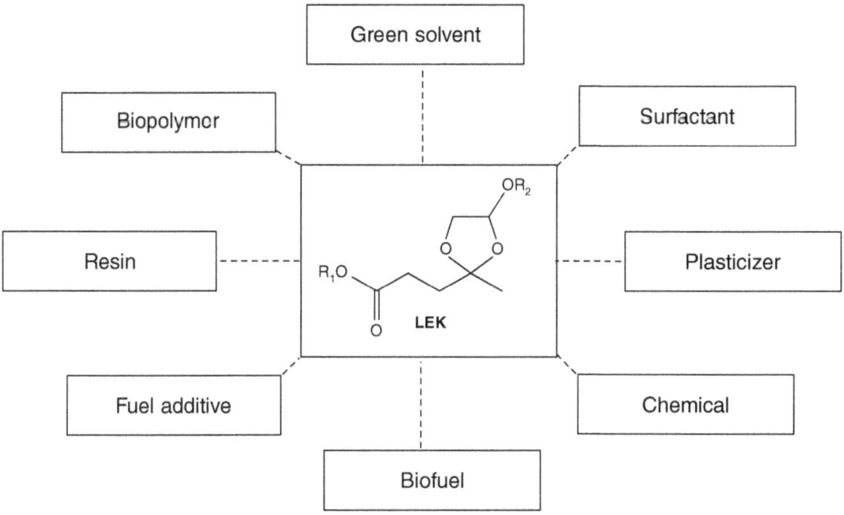

Figure 5.1 Applications of levulinic acid ester ketals (LEKs), where R_1 corresponds to alkyl groups and R_2 is hydrogen, alkyl, or acetyl groups, among others.

GLKs have been used as building block for the synthesis of plasticizers, polyols, and polymers, especially due to their biobased properties, rapid absorption, and high efficiency, low volatility, high loading, and fast-drying mixing with plastics and efficient plasticizing [18, 19]. GFBiochemical, for example, recommends the use of these types of LEKs in products for use in the automotive, textile, plastic bottles, cell phones, and footwear sectors. In the polymer industry, they are recommended for the production of polyvinyl chloride (PVC) and polyester [20].

The noncrosslinked thermoplastic polyketals have been extensively studied, especially due to their degradable nature, high O/C ratio, and production from renewable sources. These materials have been reported as promising drug delivery vehicles, finding wide biomedical applications. The combination of glycerol, levulinic acid, and diols enables the synthesis of monomers. However, the platform of monomers containing levulinic acid ketals has been further explored for the synthesis of dicarboxylic acid monomers and AB-type monomers [21]. Together with other reagents, each type of monomer polymerizes to generate different polymeric materials. Figure 5.2 shows some examples of LEK monomers used for the synthesis of polymeric materials.

Diol monomer obtained from glycerol and diethylene glycol bislevulinate ester, for example, when combined with polymeric diphenylmethane diisocyanate produces rigid polyurethane (PUs) foams. PUs are traditionally synthesized from petroleum-based chemicals, and their production from GLKs is an attractive alternative in terms of sustainability [14].

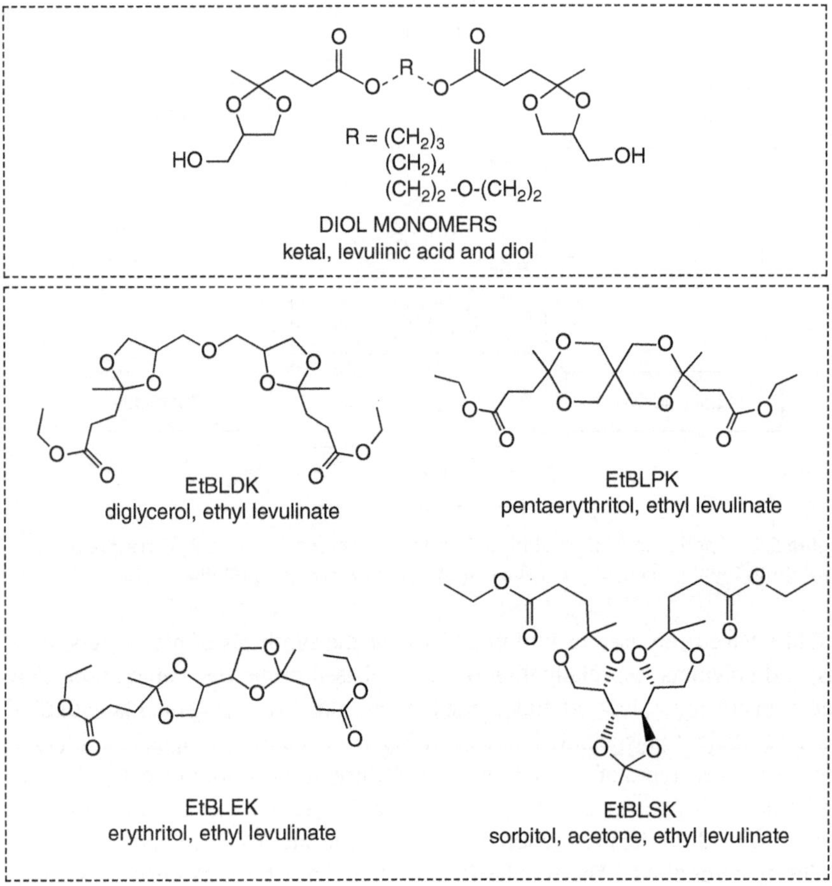

Figure 5.2 Examples of (A) diol and (B) dicarboxylic acid esters monomers based on the combination of polyalcohols and levulinic acid.

AB monomers containing a cyclo-ketal, hydroxyl, and carboxylic acid or an ester group are typically synthesized from a carboxylic acid, containing an aldehyde or a ketone functionality, and a polyalcohol [21]. Scheme 5.6 shows the formation of polyester from an AB monomer containing a cyclic ketal synthesized from glycerol and levulinic acid ester.

Amarasekara et al. were the first to report the synthesis of levulinic acid glycerol oligomers, as well as their spectroscopic characterization by NMR. Polymeric structures were obtained from the glycerol-levulinic acid system under different Lewis and Brønsted acid catalysis conditions. The low molecular weight linear oligomers showed three types of terminal units: keto, glycerol ketal, and glycerol-ester. There was no evidence for the formation of dendrimers or

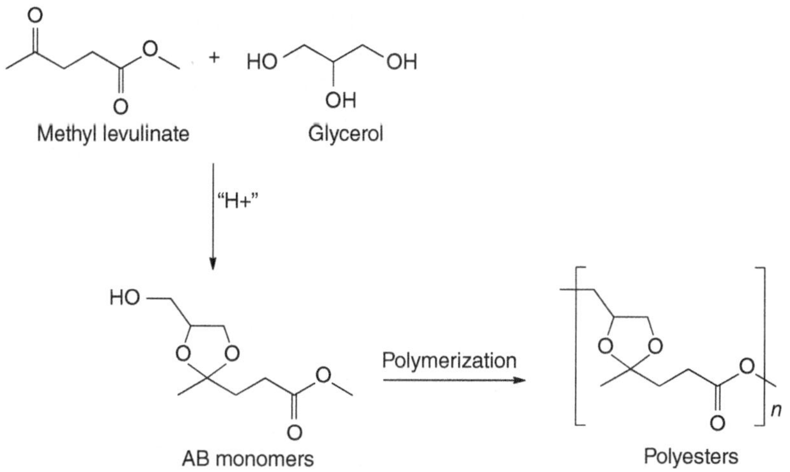

Scheme 5.6 Synthesis of polyester from an AB monomer containing a cyclic acetal, synthesized from glycerol and methyl levulinate.

branched oligomers. Authors reported the use of these materials as plasticizers and as raw materials for the synthesis of block copolyester. The highest degree of oligomerization was 9.8 for the oligomer produced in the presence of the Sb_2O_3 Lewis acid catalyst. Bronsted acid and ionic liquid catalysts were also investigated and proved to be effective, but the results were less expressive for oligomerization when compared to Lewis acid catalysts [22].

The traditional way to obtain polymeric materials, such as alkyl resins, is from glycerol and pentaerythritol monomers. However, pentaerythritol resins are more photostable when compared to resins from glycerol, since there are no secondary carbon atom in the pentaerythritol structure. Condensation between levulinic acid and pentaerythritol yields poly(levulinic acid-pentaerythritol). The reaction is similar to the reaction of levulinic acid with glycerol but affords the six-membered ring ketals [23].

GLKs in linear and branched arrangements have been investigated as plasticizers for polylactide (PLA), which is a commercial polymer synthesized from renewable resources and is considered a biodegradable alternative to other types of materials, such as polystyrene (PS) and polyethylene terephthalate (PET). However, the glass transition temperature of PLA is above room temperature, which classifies this polymer as rigid and brittle. The use of plasticizers has been shown to improve the properties of PLA. Levulinates esters containing keto terminal groups can be mentioned as those responsible for reducing the glass transition temperature (Tg) of PLA, while esters of levulinates containing ketal terminal groups present additional thermal stability [24].

Su et al. reported the synthesis of polyvinyl ketals upon the condensation of polyvinyl alcohol (PVA) and levulinic acid in the presence of p-toluenesulfonic acid as catalyst. Ketalization accounts for only 15.3% selectivity and authors associated this result with interferences from the acid functionality, since much higher selectivity to ketalization (33%) was achieved with methyl levulinate. Polyvinyl ketal is considered as a potential candidate for water degradable packaging materials [25].

5.2 Succinic Acid

5.2.1 Petrochemical and Biotechnological Routes

Two main routes for the commercial production of succinic acid are currently applied: the microbial fermentation and the chemical synthesis. Although the chemical method is cheaper than the bio-based method, it relies on petroleum-derived compounds, the price of which is extremely volatile. In addition, this route generates some by-products that are pollutants to the environment. Thus, the bio-based process is more sustainable and involves the pretreatment of the biomass and the subsequent enzymatic hydrolysis to release sugars, which in turn are used as substrates for microbial fermentation and production of succinic acid [26, 27].

Succinic acid can be industrially obtained by the catalytic hydrogenation of maleic anhydride. The petrochemical route involves the liquid phase hydrogenation of maleic anhydride to succinic anhydride at temperatures ranging from 120 to 180 °C and hydrogen pressures between 0.5 and 4.0 MPa, using palladium (Pd) or nickel (Ni) redox catalysts. The reaction is extremely exothermic with $\Delta H = -133.99\ \text{kJ mol}^{-1}$. Thus, the reactor configurations must allow an efficient heat exchange and removal [28, 29].

To increase the conversion, nitrogen is injected during the process to avoid the presence of oxygen in the reactor. The catalyst is removed upon filtration and the remaining liquid is purified by distillation at low pressure to afford succinic anhydride. The next step corresponds to hydrolysis in hot water for conversion into succinic acid (Scheme 5.7). Crystallization of succinic acid occurs upon water

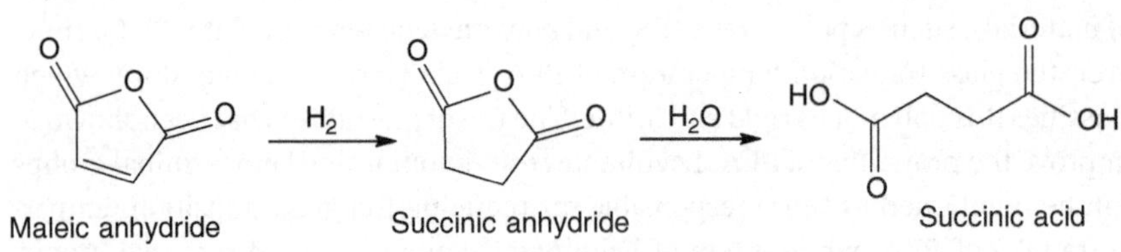

Scheme 5.7 Succinic acid production from maleic anhydride. The petrochemical route.

evaporation until saturation of the solution, followed by cooling of the system to form crystals, which are further dried [28, 29].

In a typical process, the production of 907 ton of succinic anhydride consumes 1050 Kg of maleic anhydride, 300 m^3 of hydrogen, 4500 kg of steam, 100 m^3 of cold water, 350 KW of electricity, 100 m^3 of nitrogen, and 100 m^3 of methane. The effluents to be treated are the unused catalysts, water used in the distillation tower, and the air loaded with hydrogen [29].

Succinic acid can also be obtained by hydrogenation of an aqueous solution of maleic acid in the presence of supported palladium catalysts under mild conditions. Maleic and fumaric acids are isomers and both produce succinic acid upon hydrogenation (Scheme 5.8). It is possible to achieve complete conversion of maleic acid and 100% selectivity to succinic acid at 90 °C, 5 bar, for 90 minutes, using Pd/Al_2O_3 as catalyst [30–33].

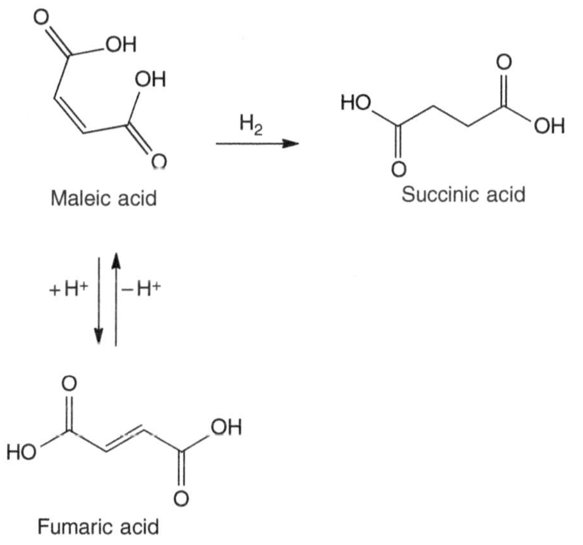

Scheme 5.8 Succinic acid production from maleic and fumaric acid.

Although the succinic acid petrochemical production routes normally present high conversions, they require catalysts based on noble metals and face environmental issues related to the use of fossil resources. Another problem that must be considered is the disposal of the unserviceable catalyst. Thus, the fermentation route has gained visibility in recent years as a promising alternative for the production of succinic acid [30].

Due to economic and environmental impacts, increased attention has been given to the production of succinic acid from biochemical processes, involving microorganisms that produce succinic acid as the main product of anaerobic fermentation of agricultural residues containing suitable fermentable C5 and C6 sugars. In addition, the fermentation to produce succinic acid consumes CO_2, contributing to the reduction of the emissions of this gas into the atmosphere [34, 35].

Succinic acid is an intermediate in the normal metabolic pathway of several microorganisms, mainly natural inhabitants of the bovine rumen. Among the microorganisms with great ability to produce succinic acid are *Anaerobiospirillum succiniciproducens*, *Actinobacillus succinogenes*, *Mannheimia succiniciproducens*, and *Escherichia coli*, and there are also studies involving some genetically modified strains. Table 5.1 shows some selected works on the production of succinic acid by fermentation [27, 47].

Many bacteria used in the production of succinic acid also produce by-products. Each system requires specific attention; for example, the bacterium *A. succiniciproducens* can present morphological changes during its growth, which can affect the yield of succinic acid. In addition, *A. succinogenes* is a bacterium capable of growing in high concentrations of sugars, which favors fermentation and enhances the production of succinic acid. On the other hand, it requires pH control due to inhibition by coproducts, such as acetic and formic acids. In

Table 5.1 Succinic acid production by different microorganisms.

Microorganism	Substrate	Fermentation	Yield ($g\,g^{-1}$)	By-products	References
A. succinogenes	Glucose	Batch	0.78	AC; FA; PA; PPA	[36]
A. succinogenes	Glucose	Fed-batch	0.87	AC; FA	[37]
A. succinogenes	Glucose	Batch	0.45	AC	[38]
A. succinogenes	Glucose	Continuous	0.85	AC; FA	[38]
A. succinogenes	Cellobiose	Batch	0.60	AC; FA; LA	[39]
A. succinogenes	Sucrose	Batch	0.58	AC; FA	[39]
A. succinogenes	Sucrose	Fed-batch	0.72	AC; FA; LA	[39]
A. succinogenes	Glucose and xylose	Batch	0.51	AC; FA; LA; PA	[40]
A. succinogenes	Duckweed	Batch	0.42	AC; FA	[41]
A. succinogenes	Cane molasses	Batch	0.76	AC; FA	[42]
M. succiniciproducens	Glucose	Fed-batch	1.27	AC; LA; PA; MA	[43]
M. succiniciproducens	Glucose	Batch	0.60	AC; FA; LA	[44]
Y. lipolytica	Glucose	Fed-batch	0.50	CA; ICA	[45]
S. cerevisiae	Glucose	Fed-batch	0.58	CA; ICA; PA	[46]

* AC: acetic acid, LA: lactic acid, FA: formic acid, PA: pyruvic acid, MA: malic acid, CA: citric acid, ICA: isocitric acid, PPA: propionic acid.

contrast, *E. coli* is more resistant to acid products than *A. succinogenes* is, but still suffers from product inhibition. Scheme 5.9 shows the biochemical pathway for the production of succinic acid from glucose using *A. succinogenes* [48–50].

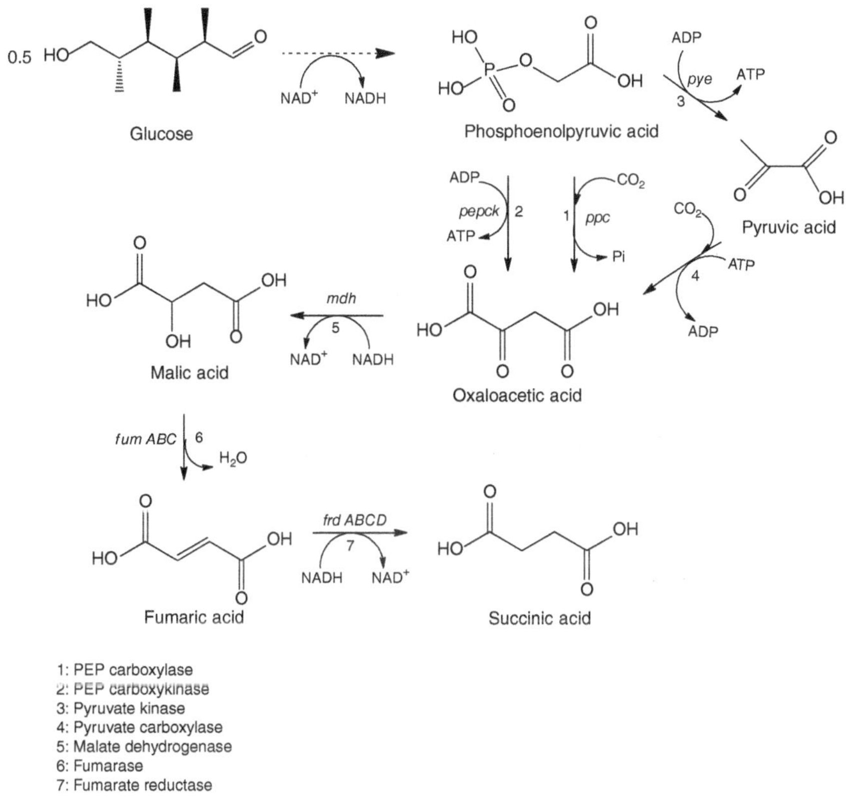

Scheme 5.9 Simplified biochemical pathway for the production of succinic acid from glucose using *Actinobacillus succinogenes*.

The production of succinic acid by the fermentative route requires milder reaction conditions. However, fermentation processes require large capacities, longer reaction times, and the use of large amounts of salt solutions. In addition, the high costs of the bioreactors and the difficulty in recovering succinic acid from the fermentation broth are some obstacles to large-scale production. In addition, efforts have been made to increase the tolerance of the microorganisms and reduce the formation of byproducts [51–53].

Consideration should be given to low pH fermentation processes or an alternative technology to minimize or avoid salt production, due to the significant costs involved. When ammonia is used during fermentation, the resulting product will be the diammonium succinate salt. Thus, sulfuric acid must be added to form

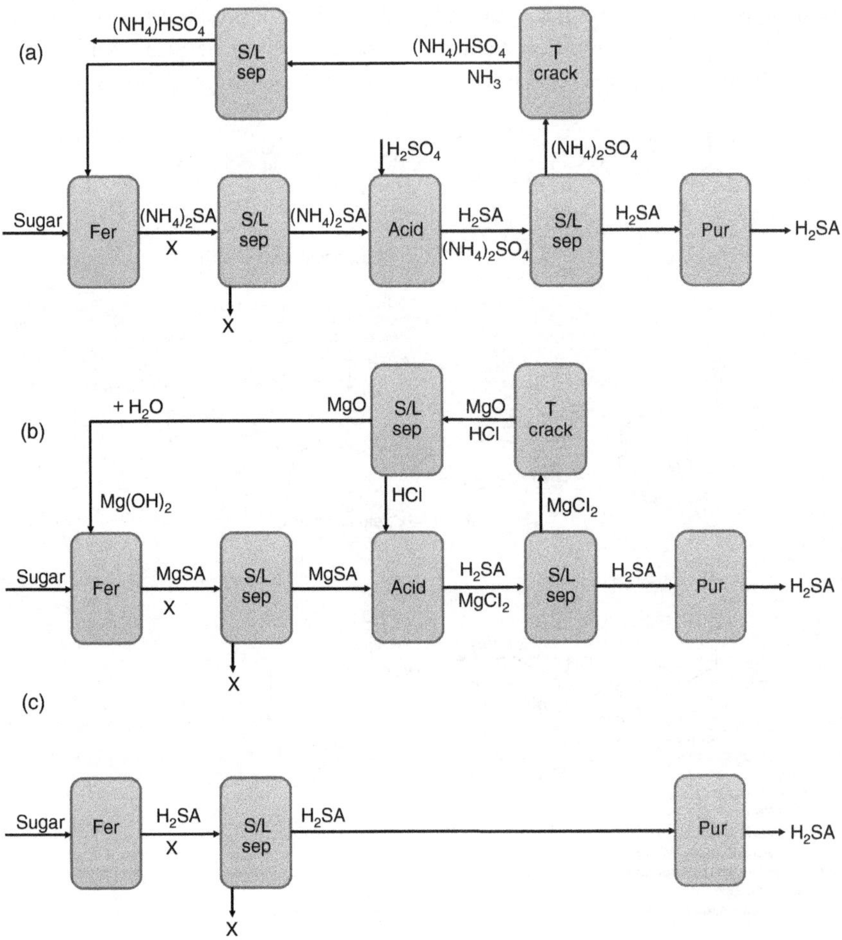

Figure 5.3 Overview of the process to produce succinic acid: (a) ammonia precipitation; (b) Mg-based process, (c) low pH fermentation. Fer: Fermentation, S/L sep: Solid liquid separation, Acid: Acidification, T crack: Cracking at high temperature, Pur: Purification, X: Biomass, AS: Succinic acid. Adapted from [54].

succinic acid and diammonium sulfate. Succinic acid can be precipitated out after concentration and further purified (Figure 5.3a). The diammonium sulfate fraction, still containing significant amounts of succinic acid, can be sold directly as a fertilizer [54].

Similarly, magnesium hydroxide is used as a pH-titrant in fermentation resulting in magnesium succinate as a product. It is necessary to add hydrochloric acid to form succinic acid and magnesium chloride. The resulting succinic acid can be separated from magnesium chloride by precipitation after concentration, and further purified. The magnesium chloride fraction, still containing minimal amounts of succinic acid, is thermally treated to afford magnesium oxide and hydrochloric acid. These two products can be recycled in the process (Figure 5.3b). The advantage of this route is that magnesium hydroxide does not affect fermentation, unlike ammonia, which can become inhibitory during fermentation [54].

The traditional method for the separation of organic acids obtained by fermentation is precipitation with calcium hydroxide. Nevertheless, this approach consumes large amounts of reagent and generates significant amounts of solid residues, accounting for about 60–70% of the cost of the final product. An advantage of the low pH fermentation process is that fewer unit operations are required, saving yield losses and investment costs (Figure 5.3c) [54, 55].

Another possibility of purification is liquid–liquid extraction using organic solvents, since succinic acid is easily extracted into the organic phase. However, the high toxicity of many solvents to the bacteria represents a major obstacle. In this sense, studies are necessary for the development of optimized and economically sustainable processes [56].

5.2.2 Levulinic to Succinic Acid

Succinic acid may be formed upon the oxidation of levulinic acid in the presence of acid catalysts and a proper oxidant, such as hydrogen peroxide, under relatively mild reaction conditions. The oxidation of LA to succinic acid occurs with the formation of various by-products and intermediates [57].

Tungstic acid catalyzes the conversion of levulinic to succinic acid using hydrogen peroxide as oxidant without the use of organic solvents. The process may be rationalized as a Baeyer–Villiger oxidation, which consists of adding an oxygen atom to convert the ketone into an ester. The peroxide attacks the LA molecule upon two competitive mechanisms to form methyl succinate or 3-acetoxypropanoic acid (Scheme 5.10) [57].

The regioselectivity of the Baeyer–Villiger oxidation with hydrogen peroxide is controlled by pH; alkaline conditions lead to the formation of the 3-acetoxypropanoic acid intermediate and latter to 3-hydroxypropanoic and acetic acids with 50% yield. Under strongly acidic conditions, using trifluoroacetic acid as a solvent, the formation of methyl succinate is favored, which subsequentially is hydrolyzed to methanol and succinic acid in 60% yield. Decarboxylation of succinic acid leads to formation of propionic acid [57].

The production of succinic acid through the vapor phase oxidative cleavage of levulinic acid with oxygen, using vanadium pentoxide as catalyst, has been reported [58]. The oxidation of levulinic acid is usually carried out using chemical oxidizing agents, such as hypohalites and nitric acid, but they are expensive. An alternative to reduce costs is the oxidation of levulinic acid with air. This approach

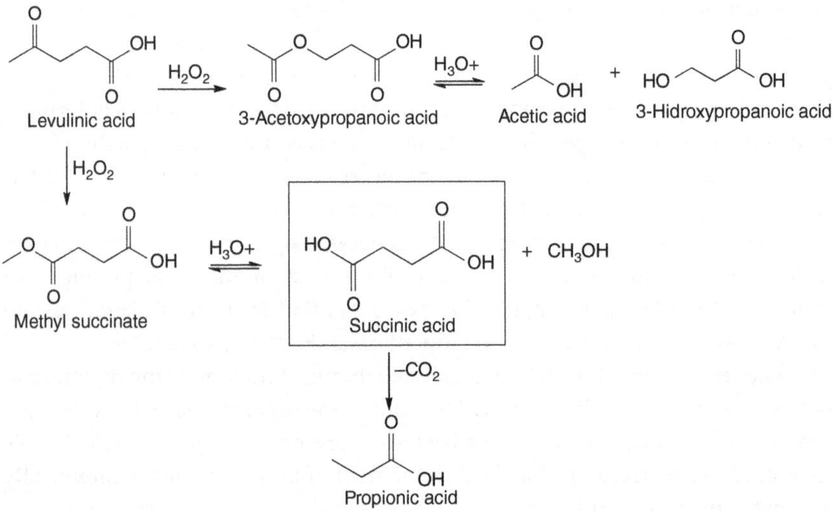

Scheme 5.10 Baeyer–Villiger oxidation of LA to succinic acid and other products [57].

has not received much attention, but it is an alternative, especially in locations with an abundance of lignocellulosic raw material.

5.2.3 Succinic Acid Main Uses

According to projections, the succinic acid market is predicted to worth US$ 182.8 million by 2023, with an annual growth rate of 6.8%. 1,4-Butanediol, one of the succinic acid derivatives, has a market forecast of US$ 14.5 billion by 2026, which will further boost the consumption and production of succinic acid [59].

There are four main existing markets for succinic acid. The first and largest is as surfactant, detergent extender, and foam agent. The secondary market refers to the use as chelating agent, which is used in electroplating to avoid corrosion of the metals. The third use refers to the food segment, where it is used as acidulant, pH modifier, flavoring agent, and antimicrobial agent. A fourth market concerns the health area, including pharmaceuticals, antibiotics, amino acids, and vitamins. Figure 5.4 illustrates the wide application of succinic acid [60].

The use of succinic acid for industrial applications represents 57.1%, whereas pharmaceutical applications correspond to 15.9%. The food sector, including beverages, represents 13.1% and the other sectors share the remaining 13.9%. All these applications have growing potential and, consequently, drive the succinic acid market [61].

Succinic acid can also be used as intermediate to products of great industrial application (Scheme 5.11). Some examples are 1,4-butanediol, an important

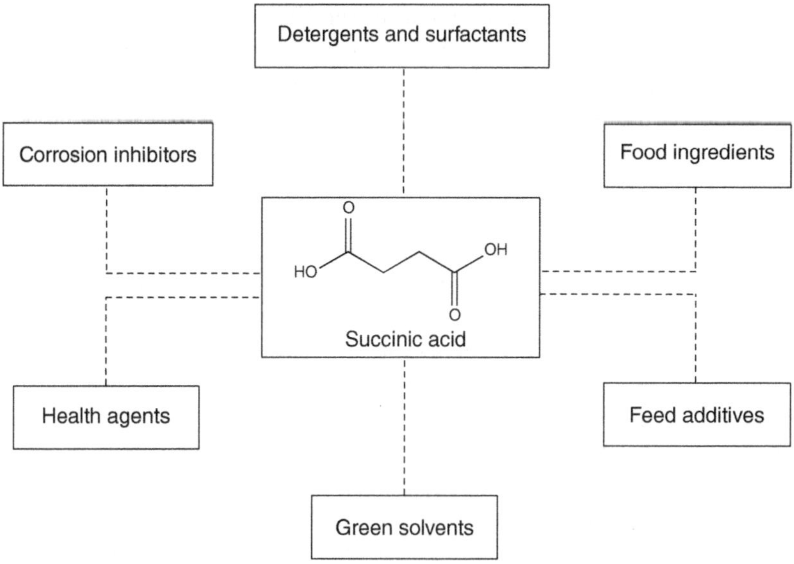

Figure 5.4 Main applications of succinic acid.

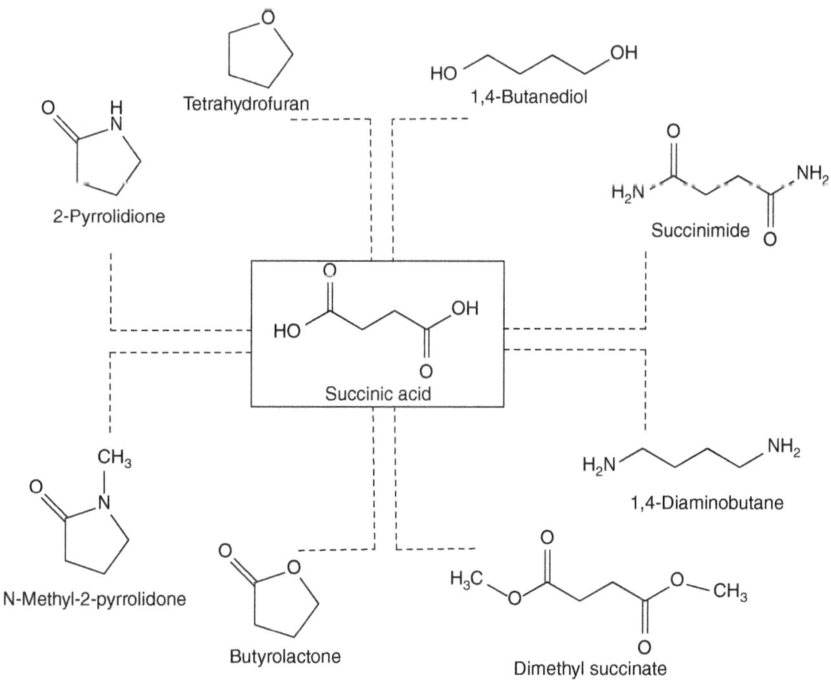

Scheme 5.11 Conversion of succinic acid into products of industrial application.

industrial solvent and raw material for resins and electrical parts; tetrahydrofuran (THF), a solvent used in the production of adhesives and printing inks; N-methylpyrrolidone, a potential substitute for methylene chloride; adipic acid, a monomer for the production of nylon. Butyrolactone, succinimide, 2-pyrrolidinone among other products can be formed from succinic acid and may find applications in polymers, resins, and paints [60, 61].

Succinic salts can be used as additives for animal nutrition, in particular, ruminants and monogastric animals, such as pigs. Sodium succinate is a flavor enhancer that can replace monosodium glutamate. Dilysine succinate is a salty flavor enhancer for low sodium foods. Succinic acid together with 1,4-butanediol is used for the production of biodegradable plastics, whereas diethyl succinate is a useful solvent for cleaning metal surfaces, whereas ethylenediamine disuccinate can replace EDTA as chelating agent [60–63].

The first industrial plant to produce succinic acid from renewable resources was built in 2009, in France, with an initial capacity of 2 kt per year. The forecast indicates that the biosuccinic acid market will reach USD 900 million by 2026. The main companies in the sector are Myriant (Canada), BioAmber (France), Succinity (Spain), and Reverdia (Italy), which together have an annual production capacity ranging from 76.6 to 86.6 Kt (Table 5.2) [64].

Table 5.2 Main companies producing bio-based succinic acid.

Company	Strains	Feedstock	Downstream	Annual Capacity (t)
BioAmber France, 2010	*E. coli*	Wheat glucose	Ammonia crystallization and electrodialysis	3 000
BioAmber-Mitsui Canada, 2014	*C. kruse*	Corn glucose	Direct crystallization at low pH	30 000
BioEnergy/Myriant United States, 2013	*E. coli*	Sugars from biomass	Ammonia precipitation	13 600
DSM/Roquette France, 2011	*E. coli*	Glucose and starch	No data	10 000–20 000
Reverdia (Roquette/DSM) Italy, 2012	*S. cerevisiae*	Starch derivatives	Direct crystallization at low pH	10 000
Succinity Spain, 2013	*B. succiniciproducens*	Glycerol or sugars	Magnesium-based process	10 000

5.3 δ-Aminolevulinic Acid (DALA) Main Routes

The obvious starting material for the preparation of δ-aminolevulinic acid (DALA), also named 5-aminolevulinic acid, is LA, but the synthetic pathway requires the formation of a C-N bond at the C5 carbon atom. Several amination methods have been successfully developed at laboratory scale, but none have been scaled up to commercial levels yet. Scheme 5.12 shows the most common approach to aminating the C5 position, which first involves the bromination of levulinic acid in alcoholic medium. The formed 5-bromolevulinic acid then reacts with potassium phthalimide, followed by acid hydrolysis to afford DALA [65]. This synthetic method presents some difficulties, such as relatively low yields and generation of large amounts of residues. Therefore, this approach is not suitable for large-scale production due to the many steps, expensive starting materials, and formation of toxic intermediates and wastes [65].

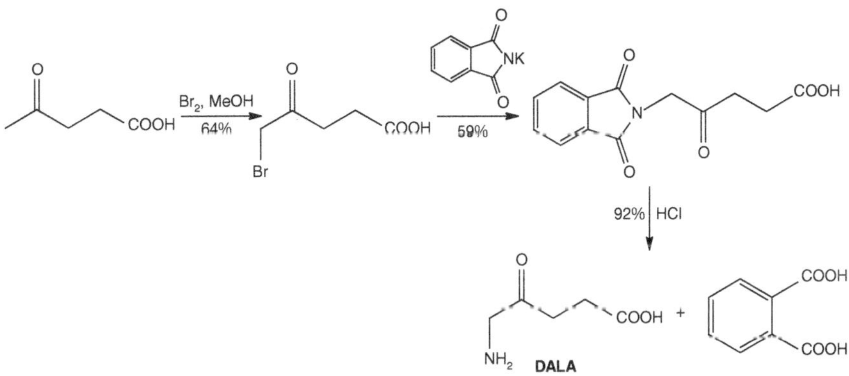

Scheme 5.12 Conventional synthesis of DALA [65].

A synthetic route that could be commercially viable is shown in Scheme 5.13. It refers to a study carried out in the National Renewable Energy Laboratory (NREL) of the United States. In this approach, levulinic acid is brominated in alcoholic medium to form methyl 5-bromolevulinate, which is further reacted with sodium diformylamide. The next step is the acid hydrolysis to yield DALA, with purity greater than 90%, together with formic acid. Each step proceeds with yields greater than 80%, and DALA was tested as herbicide showing efficiency of 85% [65].

There are many synthetic routes for the production of DALA from cyclic starting materials. For instance, DALA may be synthesized as the hydrochloride derivative from commercially available tetrahydrofurfuryl amine, which is converted to the

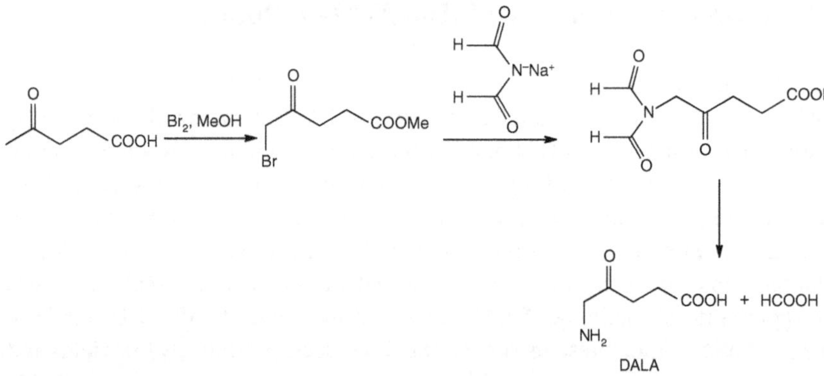

Scheme 5.13 NREL-route to the synthesis of DALA [65].

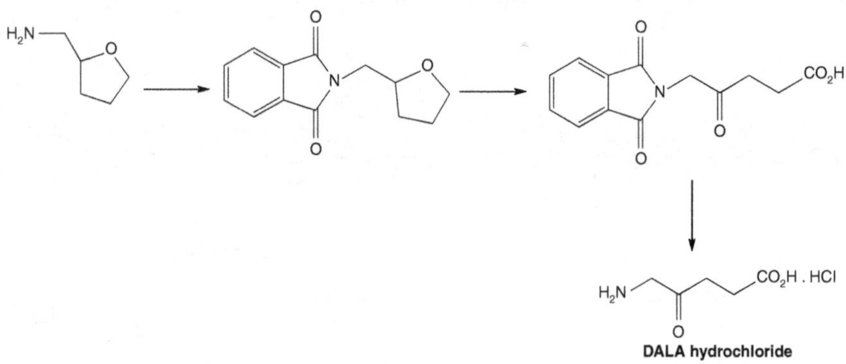

Scheme 5.14 Synthesis of DALA hydrochloride [66].

corresponding phthalimide, followed by oxidation to the corresponding ketoacid. The phthalimidyl group was then hydrolyzed with HCl forming DALA hydrochloride with an overall yield of 35.8% in 3 steps (Scheme 5.14). Although there are many studies on the synthesis of DALA, including patents, the majority of these methods proved to be arduous and involved large number of steps [66–70].

Because of the disadvantages related to chemical synthesis, microbial production of DALA has been considered as an alternative in recent years, especially with the rapid development of engineering and synthetic biology. For example, *Corynebacterium glutamicum*, *E. coli*, and *Saccharomyces cerevisiae* are the most used strains for the biosynthesis of DALA using metabolic engineering strategies [71, 72].

The production of DALA by algae and chemotrophic bacteria can be considered a less-expensive method when compared to the chemical synthesis. Nevertheless, the yield is still poor for practical implementation of this route. Table 5.3 shows

Table 5.3 Selected examples of microbial production of DALA.

Strain	Main substrates	DALA ($g\,l^{-1}$)	Ref.
Escherichia coli	Glucose	4.13	[73]
Escherichia coli	Glucose	5.25	[74]
Corynebacterium glutamicum	Glucose	2.2	[75]
Escherichia coli	Succinic acid	4.1	[76]
Saccharomyces cerevisiae	Glucose	3.6	[77]
Rhodobacter capsulatus	Glycerol, glycine, and succinic acid	8.8	[78]
Corynebacterium glutamicum	Glucose	1.79	[79]
Escherichia coli	Glucose	4.05	[80]
Escherichia coli	Glucose	3.40	[81]
Corynebacterium glutamicum	Glucose	7.54	[82]

some microorganisms and substrates used in the biotechnological production of DALA; some bacteria seem to be promising, such as prokaryotic photosynthetic bacteria and quasiphotosynthetic bacteria [83].

There are two main pathways for DALA biosynthesis in living organisms. The first is the C5 pathway, which occurs in higher plants, algae, and many bacteria; the second is the C4 pathway, which is present in mammals, birds, and photosynthetic bacteria. In the C5 pathway, glutamate is linked to the RNA to form glutamyl-tRNA, which is reduced to glutamate-1-seminalaldehyde that is rapidly converted to DALA. In the C5 pathway, DALA is formed upon the condensation of glycine and succinyl-CoA, an intermediate of the tricarboxylic acid cycle (Scheme 5.15) [84, 85].

5.3.1 δ-Aminolevulinic Acid Main Uses

DALA is a nonprotein amino acid found in living cells of bacteria, fungi, animals, and plants. It is the first common precursor in the biosynthesis of all tetrapyrroles, such as chlorophyll, heme, cytochrome, and vitamin B12. This latter compound has wide applications in medicine, nutrition, and cosmetics. Low concentrations of DALA cause many physiological effects on plant growth regulation [86–88].

Most existing pesticides are chemical preparations of high toxicity. In addition to being harmful to humans, they also present problems to the environment and the ecological systems. High doses of DALA have been used as a new photoactivated pesticide of high compatibility and environmental selectivity, because DALA is not

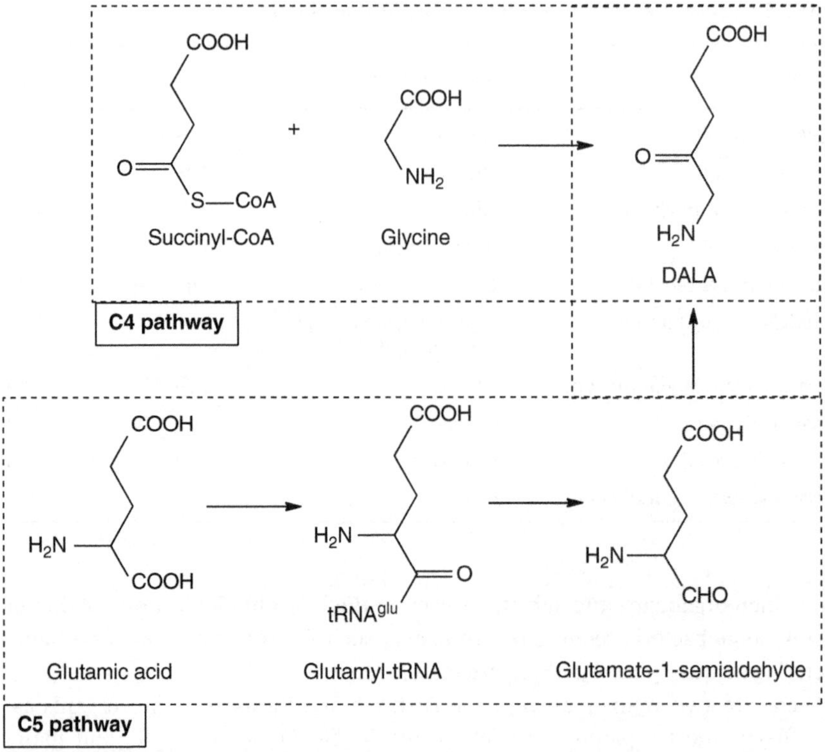

Scheme 5.15 Two main pathways for DALA biosynthesis in living organisms [84].

toxic to humans and animals, besides being biodegradable and does not generate toxic residues [88, 89].

The DALA molecule is extremely reactive and must be kept at 4 °C and protected from light. The commercially available product is sold as the hydrochloride, formed upon the protonation of the amino group. Today, DALA has limited applications because of the price around USD 67 per gram. Although it has been extensively studied, the mechanism of action and molecular basis are still not completely clear [89].

DALA is an industrial fine chemical and a biodegradable herbicide that presents high activity on dicotyledonous weeds, used as component in photodynamic therapy of cancer treatment and precancerous diseases. It is also used as a plant growth regulator and as animal food additive to improve iron status and immune response in cattle [71, 90–92].

The Biofine process focuses on optimization of the production of LA and other available products include DALA and methyl-tetrahydrofuran (MTHF) [93].

5.4 5-Methyl-*N*-Alkyl-2-Pyrrolidone Main Routes

N-heterocyclic compounds can be synthesized via reductive amination of levulinic acid. Among these, the N substituted pyrrolidone derivatives have been highlighted as a starting material and as solvent for the synthesis of pharmaceutical and agrochemical products [94]. In addition, they are considered alternative solvents for N-methyl-2-pyrrolidone, which is carcinogenic and widely used in the industry. According to Manzer, N-cyclohexyl-2-pyrrolidone is used as solvent or intermediate in many industrial applications, including the electronic industry, as photoresist removal solutions and industrial cleaners, oil well maintenance, and dyeing of fibers [95].

Many homogeneous and heterogeneous catalysts have been reported for the synthesis of N-substituted pyrrolidones from levulinic acid. Formic acid or hydrogen is usually employed as reducing reagents [94]. However, heterogeneous catalysts are preferred due to the ease of separation, reusability, integration into existing reactor equipment, and reduced waste generation. The usual synthesis conditions are 90–150 °C, 3–85 bar of H_2 pressure, and near-stoichiometric amine/levulinate molar ratio [96].

Pyrrolidones are industrially obtained from the reaction of aqueous alkylamines solutions with lactones, followed by hydrogenation in the presence of supported metal catalysts under controlled conditions of temperature and pressure [97].

The reductive amination of levulinic acid with different types of amines (aromatic and aliphatic) has been reported using heterogeneous catalyst systems [95, 98, 99]. Supported catalysts are normally used with the selected metal (Cu, Ni, Co, Fe, Ru, Rh, Pt, and Pd) being responsible for hydrogenation. Manzler et al. studied the reductive amination of levulinic acid with heptylamine or nonylamine in batch reactor, using 5% Pt/C or 5% Pd/C as catalyst in the temperature range of 150–225 °C and H_2 pressure of 3.1–8.3 bar. The yields of the pyrrolidones were within 60–82% [97].

The ketone and carboxylic functions in the levulinic acid molecule can undergo reactions with amines forming imines and amides, respectively. Therefore, there are two mechanistic possibilities for the synthesis of N-substituted-5-methyl-pyrrolidones from levulinic acid (Scheme 5.16). Each route depends on the type of catalyst, especially the type of metal [97].

In the presence of metallic catalysts (especially Pt and Rh), the first pathway normally occurs, involving the ketone functionality of levulinic acid to yield an imine. Then, there are two possibilities: (i) direct reduction of the imine to a secondary amine over the metal-catalyst, followed by acid-catalyzed cyclization to yield the desired N-substituted-5-methyl-pyrrolidone; or (ii) formation of a tautomeric imine-enamine equilibrium, which can be transformed into two distinct

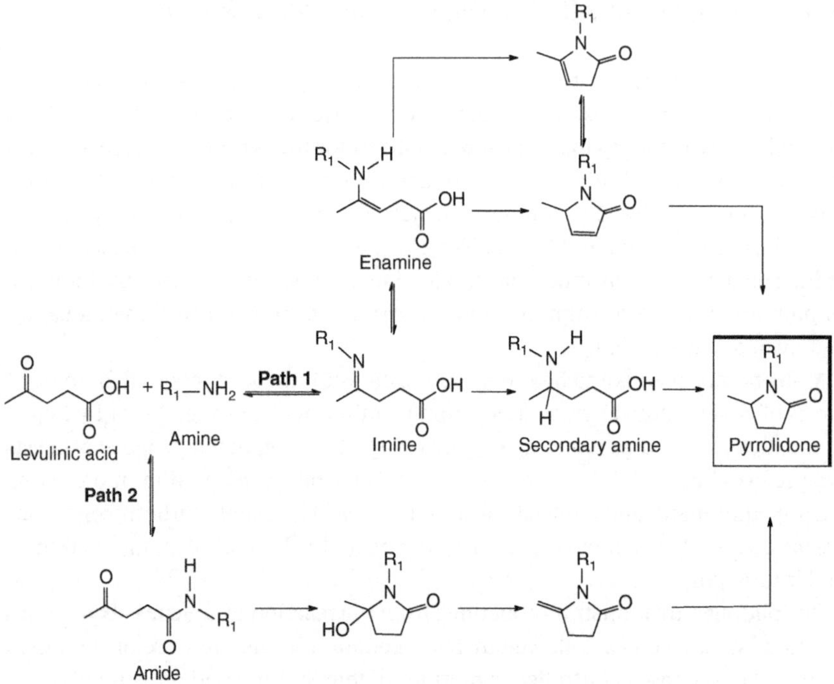

Scheme 5.16 Reductive amination of levulinic acid over supported metal catalysts [96, 97].

five-membered intermediates during cyclization. Both cyclic intermediates can be reduced to the N-substituted-5-methyl-pyrrolidones [94, 97]. It has been demonstrated that the rate-determining step of the reaction is the formation of the imine, which is accelerated by the addition of a catalyst. This observation suggests the importance of bifunctional catalysts in the direct conversion of levulinic acid to N-substituted-5-methyl-pyrrolidones [96].

The second mechanistic pathway involves the reaction of the carboxylic group in the presence of Ni-supported catalyst. First, there occurs the formation of an amide, followed by intramolecular cyclization, water elimination, and formation of a carbon double bond. Then, hydrogenation affords the N-substituted-5-methyl-pyrrolidone [97, 100].

Table 5.4 presents some selected catalysts for the reductive amination of levulinic acid to N-substituted-5-methyl-pyrrolidones. Although it is not the focus of this chapter, it is interesting to mention that conversion of organic levulinates to lactams has also been extensively reported and follows a similar reaction pathway [104, 110, 111].

Table 5.4 Reductive amination of levulinic acid with amines in the presence of heterogeneous catalysts to yield N-substituted-5-methyl-pyrrolidones.

Catalyst	H-source	H_2, bar	T, °C	Time, h	Amine	Solvent	Yield (%)	References
5%Pt-MoO_x/TiO_2	H_2	3	100	20	Aniline	—	90	[101]
5%Pt-MoO_x/TiO_2	H_2	3	100	20	Octylamine	—	99	[101]
5%Pt-MoO_x/TiO_2	H_2	3	100	20	Octylamine	Toluene	26	[101]
5%Pt-MoO_x/TiO_2	H_2	3	100	20	Octylamine	Water	4	[101]
2%Pt/P-TiO_2	H_2	1	25	3	Aniline	Met	97	[102]
2%Pt/P-TiO_2[a)]	H_2	1	25	3	Octylamine	Met	97	[102]
Pd-TiO_2	H_2	5	90	12	Octylamine	—	20	[103]
Pd-Al_2O_3	H_2	5	90	12	Octylamine	—	66	[103]
Pd-ZrO_2	H_2	5	90	12	Octylamine	—	99	[103]
Pd-ZrO_2	H_2	5	90	12	Hexylamine	—	94	[103]
Pd-ZrO_2	H_2	5	90	12	Aniline	—	83	[103]
Pd-ZrO_2	H_2	5	90	12	*p*-Chloroaniline	—	90	[103]
C $Au_{66}Pd_{34}$	H_2	1	85	12	Octylamine	—	91	[104]
Au/ZrO_2	FA	130	130	12	Aniline	Water	97	[105]
Au/ZrO_2	FA	130	130	12	Hexylamini	Water	95	[105]
Ir/SiO_2	H_2	34.5	8	100	Aniline	AcOEt	6	[106]
Ir-SiO_2-SO_3H	H_2	34.5	8	100	Aniline	AcOEt	63	[106]
Ni-Raney	FA	-	180	3	Methylamine	Water	95	[23]
Ni-Raney	FA	-	180	6	Cyclohexylamine	Ni-Raney	92	[23]
5%Pt/c-C	H_2	1	30	4	Anyline	Met	94	[107]
5%Pt/c-C	H_2	1	30	4	Octylamine	Met	97	[107]
Ni@CNT	H_2	30	130	4	Aniline	GVL	65	[100]
CNF10@Ni@CNTs	H_2	30	130	4	Aniline	GVL	68	[100]
CNF80@Ni@CNTs	H_2	30	130	4	Aniline	GVL	26	[100]
FeNi/C[b)]	H_2	85	150	25	2-phenylethylamine	2-MTHF	90	[108]
Co@Chitosan	H_2	30	130	24	Ammonium	Water	74	[109]
Zr@Chitosan	H_2	30	130	24	Ammonium	Water	3	[109]
Zr-Co@Chitosan	H_2	30	130	24	Ammonium	Water	93	[109]

a) Pt/P-TiO_2 is a porous TiO_2 nanosheet that supports Pt nanoparticles.
b) reaction carried out over continuous flow.

Metal-supported catalysts are the most used for the reductive amination of levulinic acid [101–103]. Excellent results were found for the production of pyrrolidones through the reaction of levulinic acid with amines in the presence of hydrogen on platinum supported on molybdenum and titanium oxides [101], as well as for palladium supported on zirconium oxide [103]. In general, the activity of these catalysts is attributed to the presence of strong Lewis acidity, which favors the amination of levulinic acid and hinders the direct hydrogenation-esterification side reaction [103].

Sulfonation of the support is a strategy to improve the acid properties, aiming the reductive amination of levulinic acid to pyrrolidones. Martınez et al. observed 88% yield of pyrrolidone, in 24 hours, with the use of Ir/SiO_2-SO_3H catalyst. The authors proposed a mechanistic pathway where the carboxyl group adsorbs on the sulfonic groups of the support to yield the imine, which was then hydrogenated. This mechanistic pathway was supported by theoretical calculations and experimental results. The electronegativity (μ) of the intermediate amines governs the cyclization step, mainly due to the HOMO–LUMO interaction between the amine nitrogen atom and the carboxyl group [106].

Another approach is the preparation of catalysts using atomic layer deposition. A series of porous-carbon-coated Ni catalysts (CNFx@Ni@CNTs) were prepared for the reductive amination of levulinic acid with aniline to afford 1-benzyl-5-methyl-2-pyrrolidone [100]. These catalysts were obtained with atomic layer deposition cycles (x) to control the porosity of the carbon material and the coating thickness of the Ni nanoparticles. For small x value ($x = 10$), it was observed 68% yield of the desired product, indicating that thin coatings did not block the exposure of the active metal. For high x values ($x = 80$), the yield decreased to 26%, under the same reaction conditions, suggesting that thicker coatings impair mass transfer and decrease the reaction rate. The uncoated Ni@CNTs catalyst showed 65% yield [100].

The reductive amination of LA has also been reported in continuous flow systems. Chieff et al. evaluated the reaction of levulinic acid with 2-phenylethylamine to form 2-phenylethyl-5methyl-2-pyrrolidone over carbon-supported FeNi nanoparticles. They concluded that conversion of the starting material and selectivity to the desired product increased upon using molar excess of levulinic acid. The H_2 pressure was 85 bar and 2-methyl tetrahydrofuran (2-MTHF) was used as solvent, at 0.3 ml min^{-1} of flow rate. The use of 2-MTHF is particularly convenient from a sustainability point of view, as it can be considered a biobased platform [108].

To avoid the use of noble metals and hydrogen, Amarasekara et al. evaluated the performance of Raney Ni catalyst and ammonium formate as H_2 and NH_3 sources. Levulinic acid was converted to 5-methyl-2-pyrrolidone in 94% yield using water as solvent at 180 °C [23]. Scheme 5.17 presents the proposed mechanistic pathway.

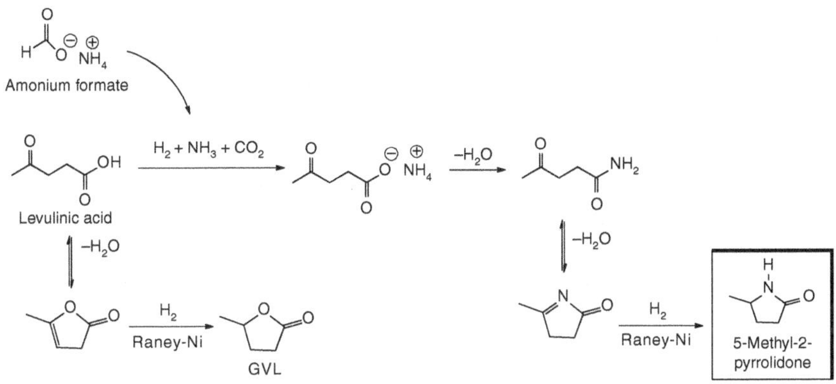

Scheme 5.17 Schematic pathway for the formation of 5-methyl-2-pyrrolidone from the reductive amination of levulinic acid in the presence of Raney Ni and ammonium formate [23].

In the presence of the same Raney Ni catalyst, N-substituted-5-methyl-pyrrolidones can be obtained from levulinic acid and a mixture of formic acid and primary amines. Levulinic acid can be converted, for example, to N-cyclohexyl-5-methyl-2-pyrrolidone in 92% yield, upon reaction with cyclohexylamine and formic acid. LA reaction with methylaniline and formic acid affords 95% yield of N-methyl-5-methyl-2-pyrrolidone [23].

Carbon materials, especially those derived from biomass, have also been employed as support in the reductive amination of levulinic acid to afford pyrrolidones. Carbon-supported Pt nanocatalyst derived from cellulose (Pt/c-C) presented 94% and 97% yields of N-substituted-5-methyl-pyrrolidones in the reaction of levulinic acid with aromatic and aliphatic amines, respectively. The high performance of the catalyst resulted from the cooperative effects of the c-C support and the Pt nanoparticles. A series of carbon and nitrogen-doped Co-M bimetallic catalysts (Co-M@Chitosan-X; M = Zr, Ni, Fe, Cu, In; X denotes the molar percentage of M) were prepared and characterized. The Co-Zr@Chitosan-X catalysts showed the presence of alloys (Co-Zr), metal-carbon bonds (Co-C, Zr-C), metal-nitrogen bonds (Co-N, Zr-N), and metal oxides (Co_3O_4, ZrO_2). Among the catalysts tested, Co-Zr@Chitosan-20 showed the best catalytic activity in the reductive amination of levulinic acid to 5-methyl-2-pyrrolidone, with 92.8% yield [109].

Some studies employed metal-free routes for the synthesis of lactams. A series of lactate-based ionic liquids (ILs) were used in the reductive amination/cyclization of ketoacids using triethoxysilane as reducing agent under mild conditions. The best result was found with 1-butyl-3-methylimidazolium lactate ([BMIm]-[Lac], which showed 80% yield for the formation of 2-phenyl-5-methyl-2-pyrrolidone from levulinic acid. Several primary amines with electron-donating or electron-withdrawing groups, including methyl, methoxy, fluorine, chlorine, bromine, and iodine, were evaluated under the established optimal conditions. The yield of the corresponding N-substituted-5-methyl-pyrrolidones ranged from 65% to 93% [112].

Bhujbal et al. developed a sustainable protocol for the reductive amination of levulinic acid to N-substituted-5-methyl-pyrrolidone, using dimethylamine borane complex (DMAB) as hydrogen source and 1-methyl piperidinium trifuoromethanesulphonate (PIL) [HmPip][OTf] as catalyst. The reaction occurs under ambient conditions and the yield of 5-methyl-1-phenyl-2-pyrrolidone was 93%. The catalyst can be recycled up to five times without significant loss of yield [113].

The reaction of levulinic acid with nitriles in the presence of H_2 is an alternative pathway for the production of N-substituted-5-methyl-pyrrolidones. Platinum, palladium, and rhodium-based catalysts showed the best results. The yield of 2-octyl-5-methyl-2-pyrrolidone was 85% on Pt-MoOx/TiO_2, 51% over Pd-MoOx/TiO_2, and 28% for Rh-MoOx/TiO_2. Upon increasing the molybdenum content from 7 to 15 wt%, the platinum catalyst showed 92% yield. Different nitriles were evaluated, and the yields of the corresponding pyrrolidones were above 70% [114]. Commercial Pd/C catalyst was efficient for the reductive amination of levulinic acid with nitriles at 80 °C, 16 bar of H_2 pressure, and THF as solvent. The yields of the corresponding pyrrolidones were up to 92% [115]. This study proposed an interesting mechanistic scheme for the formation of N-benzyl-5-methyl-2-pyrrolidone (Scheme 5.18).

The authors proposed that hydrogenation of benzonitrile to benzylamine is directly responsible for the reductive amination of levulinic acid. However, small amounts of toluene (5%), (E)-N-benzyl-1-phenylmethanimine (13%), and dibenzylamine (2%) were also observed, which poses complexity to the process due to formation of by-products. The formation of (E)-N-benzyl-1-phenylmethanimine is reversible, and equilibrium is shifted upon consumption of benzylamine in the reductive amination of levulinic acid.

Other possible secondary reactions are: (i) condensation of levulinic acid with NH_3 formed from benzonitrile to obtain 5-methyl-2-pyrrolidone; and (ii) hydrogenation of levulinic acid over Pd/C catalyst to yield γ-valerolactone. Finally, the Schiff base generated from the reductive amination of levulinic acid with benzylamine follows the mechanistic pathways already established, with formation of 2-benzyl-5-methyl-2-pyrrolidone [115].

Considering the formation of N-substituted-5-methyl-pyrrolidones from levulinic acid, there are two other possibilities: one involving aldehydes/ketones and ammonia as reagents, and the second involving nitro compounds, both in the presence of hydrogen. In the first case, the formed amine reacts with levulinic acid in the presence of hydrogen to form N-substituted-5-methyl-pyrrolidones [113]. In the second possibility, levulinic acid reacts with the nitro compound to form

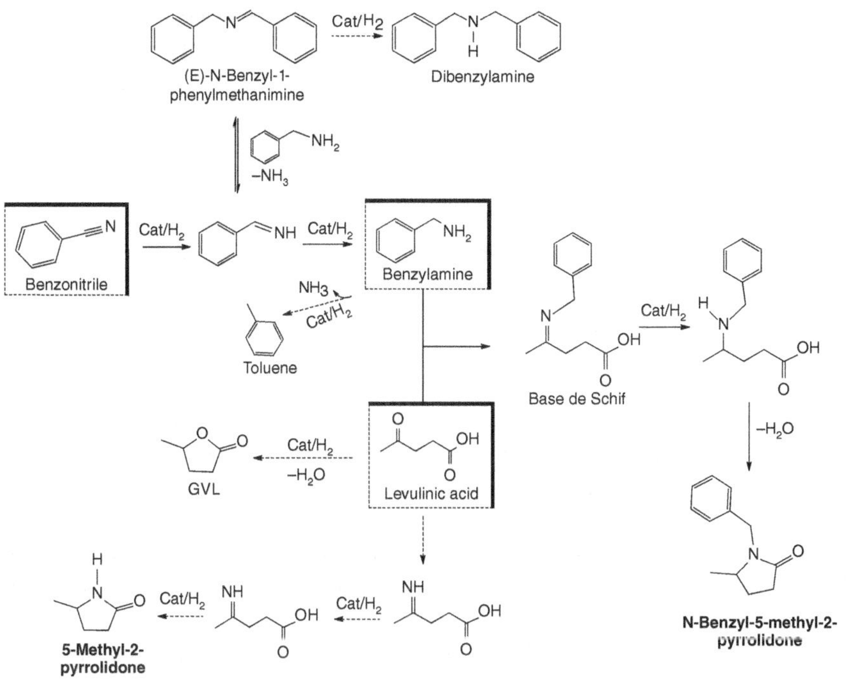

Scheme 5.18 Suggested reaction pathway for the reductive amination of levulinic acid with benzonitrile [115].

N-substituted 5-methyl-2-pyrrolidone. Different catalysts and conditions may be employed. Metal catalysts (palladium, ruthenium, rhenium, rhodium, iridium, platinum, nickel, cobalt, copper, iron, osmium) supported on high-specific area materials (carbon, titanium, or aluminum oxides) together with metal promoters (zinc, copper, gold, silver) showed good results. The nitro compound—LA molar ratios ranged from 0.3 : 1 to 5 : 1, and the temperature is kept between 75 C and 200 °C, with hydrogen pressures from 13 to 76 bar [98].

5.4.1 5-Methyl-N-Alkyl-2-Pyrrolidone Main Uses

N-substituted-5-methyl-pyrrolidones can be used as solvents or intermediates in many industrial applications in the electronic sector, industrial cleaning products, maintenance of oil/gas wells, fiber dyeing, agrochemicals, and pharmaceuticals [98, 116]. N-[2-Hydroxyethyl-]2-pyrrolidone is widely used in industrial cleaning, printing inks, and as gasoline and oil additives, whereas N-octyl-2-pyrrolidone is used in the manufacture of agricultural products, such as detergents and dispersants [117]. The N-substituted-5-methyl-2-pyrrolidones have been considered an alternative substitute for the N-substituted pyrrolidones, especially

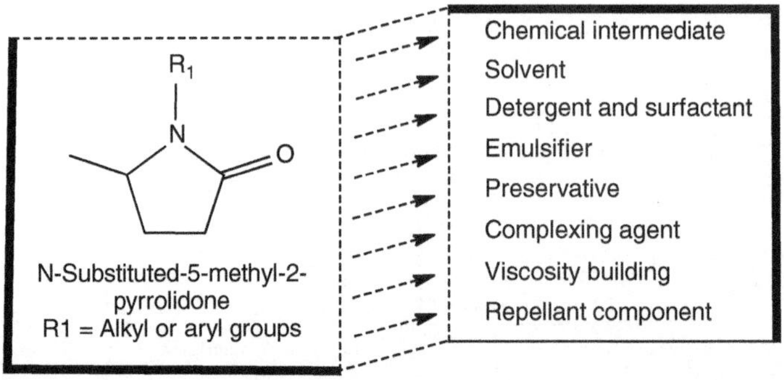

Figure 5.5 Application of N-substituted-5-methyl-2-pyrrolidone.

because of their similar chemical characteristics [98]. However, depending on the size of the alkyl chains in the pyrrolidone structure, there are important changes in the properties of the molecules. For example, when the alkyl chains are lower than eight carbon atoms, the pyrrolidone derivatives show properties of aprotic solvent and low toxicity. When the alkyl chain is around 14, the pyrrolidone is employed as surfactant. Figure 5.5 shows different applications of N-substituted-5-methyl-2-pyrrolidones.

Due to their solvent, surfactant, and complexing properties, N-substituted-5-methyl-2-pyrrolidones have been useful, especially in the manufacture/formulation of pharmaceutical, agrochemical, and cleaning products. The agrochemical sector includes the manufacture of pesticides (herbicides, insecticides, fungicides, among others), plant growth regulators, and repellants. The latter products can be used in liquid or aerosol formulations for dermal application in humans and animals, such as mosquito and tick repellents, respectively.

In the pharmaceutical industry, N-substituted-5-methyl-2-pyrrolidones have been used on various topical formulations (ointments, lotions, gels, sprays, aerosols, among others) in humans and animals, as they increase the absorption of many drugs by the skin. Its transdermal uses can be exploited for the administration of antimicrobial, hormone, or anti-inflammatory medicaments [98, 116]. Furthermore, the pyrrolidone core is present in several drug structures (Figure 5.6). The exemplified drugs are correlated with nootropic, neuroprotective, posttraumatic, and antiepileptic effects [118].

Polyvinylpyrrolidone (Povidone or PVP) has been used as binder, coating, and disintegrant to stabilize tablets. PVP has also been used in other areas, such as cosmetics, food, adhesives, polymers, and textiles [119].

In the cleaning sector, various types of formulations include N-dodecyl-5-methyl-2-pyrrolidone, especially degreasing formulations. In addition, this compound is used as liquid remover, disinfectant, and antimicrobial agents. Other formulations that employ pyrrolidones include products for automotive paint protection and inkjet component [98].

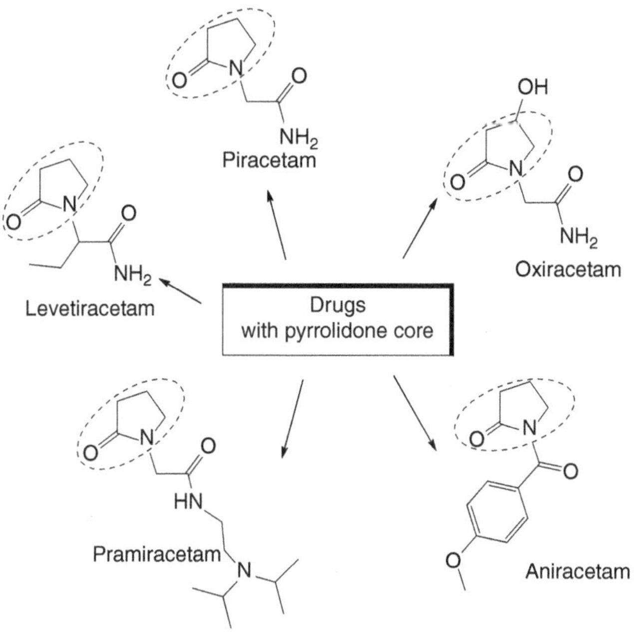

Figure 5.6 Examples of drugs with the pyrrolidone core.

5.5 Diphenolic Acid Main Routes

Diphenolic acid (DPA), also denominated 4,4-bis(4-hydroxyphenyl) pentanoic acid, is an important derivative of LA. Table 5.5 summarizes some selected properties of DPA.

DPA is produced through the condensation of two moles of phenol with one mol of levulinic acid (LA) (Scheme 5.19). Due to the structural similarities, diphenolic acid is considered a good candidate to replace bisphenol A (BPA), the main raw material of epoxy resins. However, the lower cost of BPA has impaired the widespread application of DPA. The high cost of levulinic acid contributes to the price of DPA being around USD 6 per kilogram. With the increasing production of LA from biomass, the price may become more competitive (USD 2.4 per kg), which allows the gradual replacement of BPA by DPA [93]. BPA is produced from fossil resources, using acetone and phenol. In addition, it presents severe hormonal impact on the body, as it plays the role of estrogen. BPA has been linked to an increased risk of lowering fertility and cancer [120]. The Biofine process offers LA at low cost, and in this way, DPA can be considered a precursor for renewable-based polymers [121].

A major advantage of this reaction is the use of water as solvent. It is also possible to replace phenol by other naturally occurring phenolic compounds, such

Table 5.5 Selected properties of DPA.

Physical properties	DPA
Formula	$C_{17}H_{18}O_4$
Mol wt ($g\,mol^{-1}$)	286.33
Flash point (°C)	208
Melting point (°C)	167–170 °C
Boiling point (°C)	507 °C
Density ($g\,ml^{-1}$)	1.30–1.32
Refractive index	1.675
Potential market demand ($mt\,y^{-1}$)	10–20
Solubility	Slightly soluble in water, soluble in acetic acid, acetone, ethanol, isopropanol, methyl ethyl ketone, and insoluble in benzene

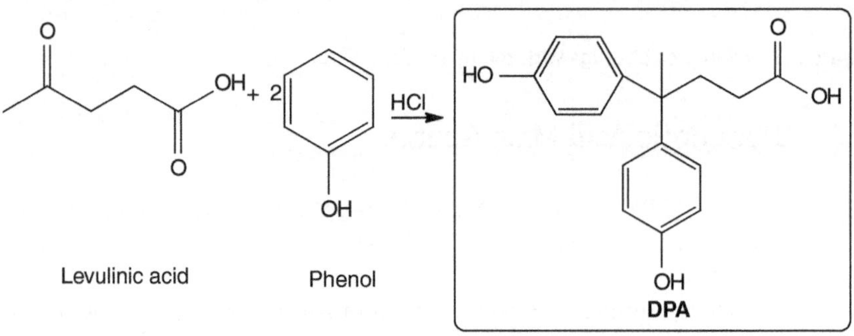

Scheme 5.19 General synthetic route of DPA.

as catechol, resorcinol, guaiacol, and m-cresol. Catechol and resorcinol showed satisfactory yields; 55 and 65%, respectively [122].

The synthesis of DPA is traditionally catalyzed by strong mineral acids, such as concentrated HCl or H_2SO_4, with molar ratio of phenol to levulinic acid in the range of 2.5–4, and reaction times of up to 60 hours. Under these conditions, the maximum yield of DPA is 85%, based on the levulinic acid conversion [121]. Nevertheless, there are some disadvantages, such as corrosivity, difficult separation, and expensive waste treatment. In recent decades, environmental and economic issues have stimulated the use of heterogeneous solid acid catalysts for the production of DPA, replacing toxic and corrosive mineral acid catalysts.

Zeolites have drawn attention to the synthesis of DPA, but showed limitations when large reactants and/or product molecules are involved [123]. The synthesis of DPA has been investigated using Keggin-type HPAs [6]. While HPAs provide strong Brønsted acidity, in the bulk form, they present low efficiency due to low-surface area [7, 8]. Guo et al. studied the catalytic activity of as-prepared $H_3PW_{12}O_{40}/SiO_2$ and compared it with HCl, $H_3PW_{12}O_{40}$, $H_3PW_{12}O_{40}/SiO_2$-15.4. Under an excess of phenol and solventless conditions, all acids were efficient. In all cases, the two structural isomers of DPA were formed. The yield of DPA increased with increasing the loading of $H_3PW_{12}O_{40}/SiO_2$-C-15.7. The reactions were carried out at 80, 100, and 120 °C, and the overall DPA yield increased with reaction temperature, although the selectivity for DPA decreased. Increasing the phenol to LA molar ratio from 2 : 1 to 5 : 1 led to an increase in DPA yield after 6 h, using $H_3PW_{12}O_{40}/SiO_2$-E-14.8 as catalyst. Notwithstanding, the homogeneous HCl or $H_3PW_{12}O_{40}$ catalytic systems showed higher LA conversion compared to the solid materials.

The synthesis of DPA was studied over Wells–Dawson-type heteropolyacid, substituted with Cs, at 150 °C, obtaining 85% yield of product [124]. Sulfonated hyperbranched poly(arylene oxindoles) were prepared and catalyzed the condensation of phenol and LA with 1:1 molar ratio of thiols as additive [125].

SO_3H-based ionic liquids were used in the synthesis of diphenolic acid from the condensation of phenol and levulinic acid. Both, p,p′- and o,p′ DPA isomers were observed (Scheme 5.20). The conversion was above 90%, with 100% of selectivity to p,p′-DPA, at 60 °C and 24 hours, using [BSMim]HSO_4 [124].

5.5.1 Diphenolic Levulinic Acid Main Uses

DPA is commonly used as intermediate and monomer in the chemical sector. However, it has also found applications in several other areas. DPA is a potential candidate to replace bisphenol-A in cosmetics, surfactants, plasticizers, paint formulations, textile additives, coatings, and lubricating oil additives [126]. Figure 5.7 presents some potential applications.

DPA/BPA has been used in combination with polycarbonate and polyacrylate to partially replace BPA formulations. Brominated diphenolic acid can be used as an environmentally acceptable marine coating, whereas dibrominated diphenolic acid is a fire retardant [121].

Phenolic acid derivatives are considered excellent monomers to replace bisphenol-A. Its self-curing capacity favors the production of epoxy resins, avoiding the use of crosslinking agents that may be toxic and harmful to the environment. Because epichlorohydrin can be produced from glycerol, a by-product of biodiesel production, the cured epoxy resin may be considered fully renewable. As the main use of bisphenol-A is in epoxy resins, Caretti et al. reacted diphenolic

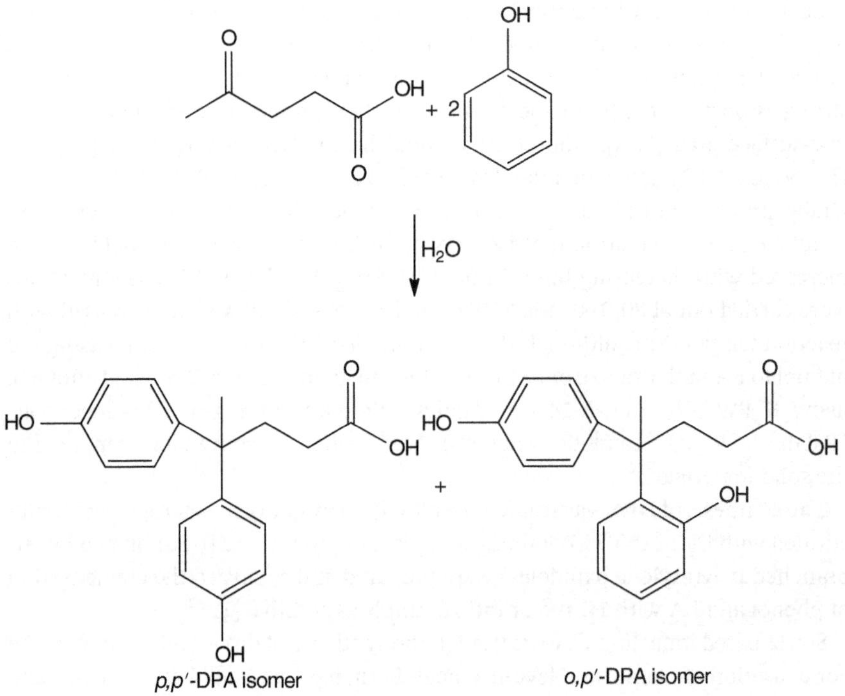

Scheme 5.20 Condensation of LA and phenol to afford the two DPA isomers.

acid derivatives with epichlorohydrin to assess their ability to form glycidyl ethers that would be cured into a resin. DPA derivatives of resorcinol and catechol showed interesting yields. Furthermore, as they have five functional groups, there was no need to use a crosslinking agent [122]. Preliminary studies showed that the obtained resin presented good properties, with excellent pencil hardness for the catechol/DPA derivative. Thermal stability is comparable to other epoxy resins.

The syntheses of DPA-based benzoxazine and DPA ester derivatives, based on benzoxazine, were reported [127]. The presence of a carboxylic acid in the DPA monomer should lower the temperature needed to complete the curing of these benzoxazines.

Bisphenol-A is toxic and produced from nonrenewable sources. Thus, it is not sustainable for the synthesis of epoxy resins. On the other hand, DPA is produced from renewable LA, but its carboxyl group limits its applications in biobased thermosets. To remedy this inconvenience, the carboxyl groups were transformed into ester functionality [128, 129]. The resulting ester groups were, however, partially hydrolyzed in alkaline medium, forming by-products that led to low-purity epoxides [130]. Epoxy resins mixed with a characteristic hardener agent represent most of the thermoset polymers, with diglycidyl ether bisphenol-A (DGEBA) being the

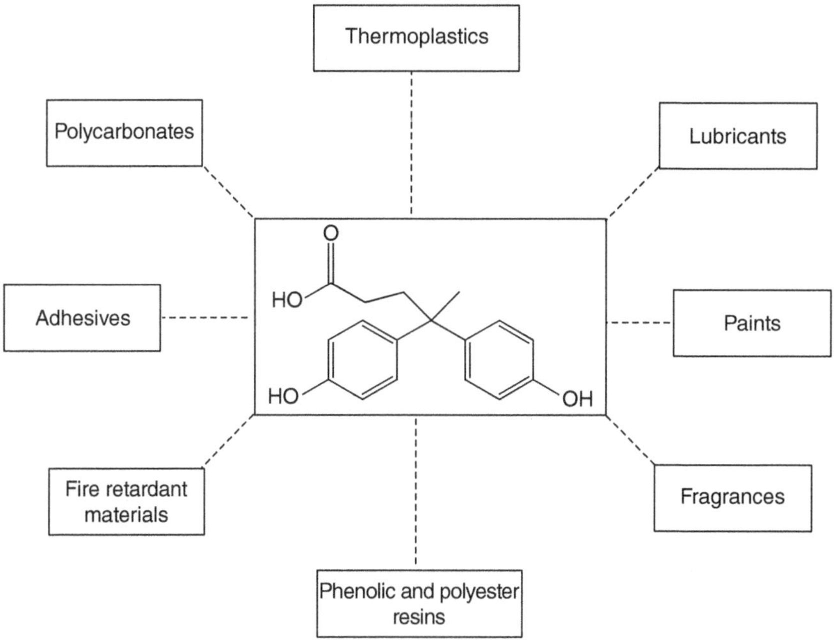

Figure 5.7 Some potential applications of DPA.

most used. These resins are used as adhesives and coatings. DGEBA is a liquid epoxy resin that can be processed at low temperatures and provides good wettability to the reinforcement fillers.

Qian et al. studied a new amidation route to adapt DPA to produce epoxy monomers that feature epoxy resins of high acting. They synthesized three BPA analogs upon amidation of DPA with different amines, such as diphenolic ethylamide (DEA), diphenolic butylamide (DBA), and diphenolic hexamine (DHA) (Scheme 5.21) [131]. This process was inspired by the industrial production of nylon-66 [132]. Diphenolic amides showed low toxicity when compared to BPA. The amidation of DPA can easily occur because amines are more reactive than the alcohol analogs [131]. The improved thermomechanical properties of the cured epoxy resins were attributed to increased crosslink density and increased hydrogen bonding by the amide groups. In addition, DPA-derived epoxy resins have great potential in the preparation of biobased composites, as they act as excellent interfaces between the resin and the cotton fibers [131].

The CN103058831A patent discloses the synthesis of diphenolic acid derivatives, such as bisphenolic alcohol, aliphatic amine/diphenolic acid aromatic amine amidate, diphenolic acid epoxy resin, and diphenolic acid glycidyl ether epoxy resin. The method of diphenolic acid derivatives preparation presents high productivity,

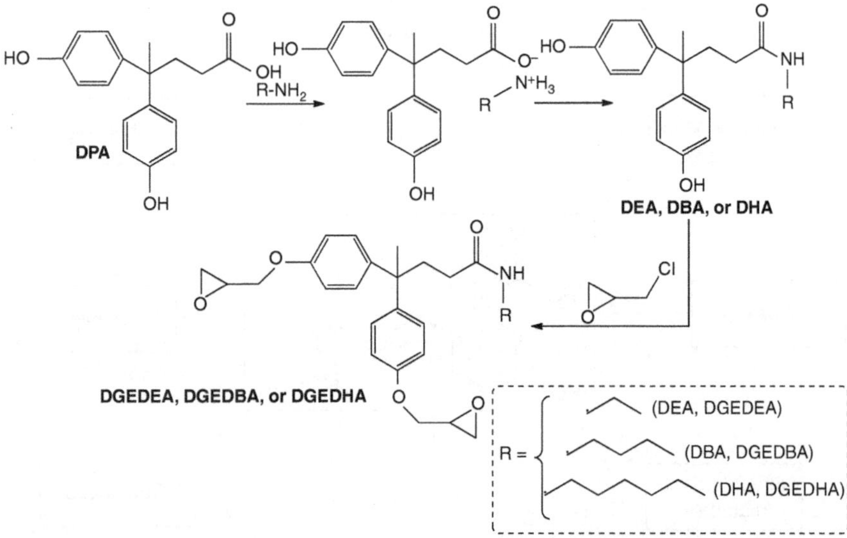

Scheme 5.21 Synthesis of BPA analogs upon the amidation of DPA [131].

Table 5.6 Industrial application of DPA.

Diphenolic Acid Applications	Patent Reference	Reference
Thermoplastics	BASF A.-G., 1998	[135]
Adhesives	S.C. Johnson & Son, Inc., 1959	[136]
Lubricants	Texaco, 1995	[137]
Paints	Minnesota Mining and Mfg. Co., 1977	[138]
Electronics	Shin-Etsu Chemical Industry, 1998	[139]
Medicinal	Abbott Laboratories, 1997	[140]
Polycarbonates	General Electric, 1974	[141]
Printing/inks	Xerox, 1997	[142]
Fragrances	Givaudan-Roure (International) S.A., 1997	[143]
Fenolic and polyester resins	Ciba-Geigy A.-G., 1986	[144]
Fire retardant materials	General Electric Co, 1991	[145]

easy separation, and simple reaction conditions. It can be used to synthesize harmless epoxy resins and polycarbonates [133]. The CN102701955B patent used acidic ionic liquid for the synthesis of diphenolic acid and/or diphenolic acid ester [134].

DPA is widely used in polymers but finds applications in other areas, as shown in Table 5.6.

Chemical modifications to the DPA functional groups may provide more flexibility for the wide application of this compound. The use of DPA as a substitute for bisphenol-A and the study of the properties of the resulting polymers will provide new sustainable opportunities for the chemical industry, making DPA a "green" candidate for a variety of applications.

5.6 Conclusion

Several derivatives are formed from reactions on the carbonyl group of levulinic acid. Although each product has its peculiarities, all have great industrial applications. Among them is succinic acid, which can find many applications in different sectors.

DALA is another important derivative, but the synthesis methods still present some difficulties, such as relatively low yields, generation of large amounts of residues, and many steps.

Ketals derived from levulinic acid can be considered green and sustainable, being good building blocks for the synthesis of several products, especially for uses as plasticizers and polymers. The glycerol levunic ester ketal (GLEK) is commercially produced and is a completely renewable molecule.

The 5-methyl N-substituted pyrrolidones find application as solvents, and chemical intermediates for the production of different products, especially those destined to the pharmaceutical and agrochemical sectors.

Diphenolic acid (DALA) is a potential substitute for BPA in the production of polycarbonates, epoxy resins, and other polymers, enabling new sustainable opportunities for the chemical industry.

References

1 Tulchinsky, M.L. and Briggs, J.R. (2016). One-pot synthesis of alkyl 4-alkoxypentanoates by esterification and reductive etherification of levulinic acid in alcoholic solutions. *ACS Sustainable Chem. Eng.* 4 (8): 4089–4093.

2 Antonetti, C., Licursi, D., Fulignati, S. et al. (2016). New frontiers in the catalytic synthesis of levulinic acid: from sugars to raw and waste biomass as starting feedstock. *Catalysts* 6 (12): 196.

3 Mullen, B.D., Badarinarayana, V., Santos-Martinez, M., and Selifonov, S. (2010). Catalytic selectivity of ketalization versus transesterification. *Top. Catal.* 53 (15–18): 1235–1240.

4 Freitas, F.A., Licursi, D., Lachter, E.R. et al. (2016). Heterogeneous catalysis for the ketalisation of ethyl levulinate with 1,2-dodecanediol: opening the way to a new class of bio-degradable surfactants. *Catal. Commun.* 73: 84–87.

5 Umrigar, V., Chakraborty, M., and Parikh, P. (2019). Esterification and ketalization of levulinic acid with desilicated zeolite β and pseudo-homogeneous model for reaction kinetics. *Int. J. Chem. Kinet.* 51 (4): 299–308.

6 Park, S.L. and Seli, M.N.U.S. (2009). Ketal compounds from polyols and oxocarboxylates. WO 2009/032905 Al. Issued 2009.

7 Sergey Selifonov, Scott D. Rothstein, Brian D. Mullen (2010). (12) Patent Application Publication (10) Pub. No.: US 2010/0292491 A1. **1** (19), issued 9 April 2010.

8 Selifonov, S., Rothstein, S.D., and Mullen, B.D. (2013). Method of making ketals and acetals. United States Patent, 2 (12).

9 Doll, K.M. and Erhan, S.Z. (2008). Synthesis of cyclic acetals (ketals) from oleochemicals using a solvent free method. *Green Chem.* 10 (6): 712–771.

10 Amarasekara, A.S. and Animashaun, M.A. (2016). Acid catalyzed competitive esterification and ketalization of levulinic acid with 1,2 and 1,3-diols: the effect of heterogeneous and homogeneous catalysts. *Catal. Lett.* 146 (9): 1819–1824.

11 Konwar, L.J., Samikannu, A., Mäki-Arvela, P. et al. (2018). Lignosulfonate-based macro/mesoporous solid protonic acids for acetalization of glycerol to bio-additives. *Appl. Catal., B Environ.* 220: 314–323.

12 Valley, G. and Data, R.U.S.A. (2013) Adducts of levulinic derivatives with epoxdzed fatty acid esters and uses thereof. (12) Patent Application Publication (10) Pub. No.: US 2013/0233204 A1. 1 (19), issued 12 September 2013.

13 Stepan, E., Enascuta, C.E., Oprescu, E.E. et al. (2018) A versatile method for obtaining new oxygenated fuel components from biomass. *Ind. Crops Prod.* 113: 288–297.

14 Li, P., Xiao, Z., Chang, C. et al. (2020). Efficient synthesis of biobased glycerol levulinate ketal and its application for rigid polyurethane foam production. *Ind. Eng. Chem. Res.* 59 (39): 17520–17528.

15 Yontz, I.D.J., Rieth, L.R., Us, M.N., Morante, N., Us, N.Y., Palefsky, I., and Us, N.J. (2017) Personal care formulations containing alkyl ketal esters and methods of manufacture; Patent number: US 9594886B2.

16 Valley, G. and Data, R.U.S.A. (2012). Glycerol levulinate ketals and their uses; Patent number: US 20120021962A1.

17 Valley, G. (2014). Alkyl ketal esters as dispersants and slip agents for particulate solids, methods of manufacture and uses thereof.

18 Valley, G. (2012). Glycerol levulinate ketal and their use in the manufacture of polyurethanes, and polyurethanes formed therefrom, issued 02 August 2012.

19 Larson, G., Holmes, C., Smith, B., and Wilson, D.D. (2013). Stabilized levulinic ester ketals. (12) Patent Application Publication (10) Pub. No.: US 2013/0344194A1. **1** (19), 2–6.

20 Wypych, A. (2017). Levulinic acid and its derivatives. In: *Datab. Plast.*, 2e, 337–340. ChemTecToronto ISBN 978-1-895198-96-6.

21 Hufendiek, A., Lingier, S., and Du Prez, F.E. (2019). Thermoplastic polyacetals: chemistry from the past for a sustainable future? *Polym. Chem.* 10 (1): 9–33.

22 Amarasekara, A.S. and Hawkins, S.A. (2011). Synthesis of levulinic acid-glycerol ketal-ester oligomers and structural characterization using NMR spectroscopy. *Eur. Polym. J.* 47 (12): 2451–2457.

23 Amarasekara, A.S. and Lawrence, Y.M. (2018). Raney-Ni catalyzed conversion of levulinic acid to 5-methyl-2- pyrrolidone using ammonium formate as the H and N source. *Tetrahedron Lett.* 59 (19): 1832–1835.

24 Xuan, W., Hakkarainen, M., and Odelius, K. (2019). Levulinic acid as a versatile building block for plasticizer design. *ACS Sustainable Chem. Eng.* 7 (14): 12552–12562.

25 Su, Y.K., Coxwell, C.M., Shen, S., and Miller, S.A. (2021). Polyvinyl alcohol modification with sustainable ketones. *Polym. Chem.* 12 (34): 4961–4973.

26 Chen, J., Yang, S., Alam, M.A. et al. (2021). Novel biorefining method for succinic acid processed from sugarcane bagasse. *Bioresour. Technol.* 324: 124615.

27 Yang, Q., Wu, M., Dai, Z. et al. (2020). Comprehensive investigation of succinic acid production by *Actinobacillus succinogenes*: a promising native succinic acid producer. *Biofuels, Bioprod. Biorefin.* 14 (5): 950–964.

28 Kim, J.H., Choi, J.H., Kim, J.C. et al. (2021). Production of succinic acid from liquid hot water hydrolysate derived from *Quercus mongolica*. *Biomass Bioenergy* 150: 106103.

29 Fumagalli, C. (2006). Succinic acid and succinic anhydride. In: *Kirk-Othmer Encyclopedia of Chemical Technology*, 15.

30 Byun, M.Y., Kim, J.S., Baek, J.H. et al. (2019). Liquid-phase hydrogenation of maleic acid over Pd/Al_2O_3 catalysts prepared via deposition–precipitation method. *Energies* 12 (2): 3–10.

31 Kulagina, M.A., Gerasimov, E.Y., Kardash, T.Y. et al. (2015). A universal method to form Pd nanoparticles on low-surface-area inorganic powders and their support-dependent catalytic activity in hydrogenation of maleic acid. *Catal. Today* 246: 72–80.

32 Ruiz, P., Crine, M., Germain, A., and L'Homme, G. (1982). Influence of the reactional system on the irrigation rate in trickle-bed reactors. *Annu. Meet. – Am. Inst. Chem. Eng.*, 15–36.

33 Kulagina, M.A., Simonov, P.A., Gerasimov, E.Y. et al. (2017). To the nature of the support effect in palladium-catalyzed aqueous-phase hydrogenation of maleic acid. *Colloids Surfaces A Physicochem. Eng. Asp.* 526: 29–39.

34 Rezaei, M.N., Aslankoohi, E., Verstrepen, K.J., and Courtin, C.M. (2015). Contribution of the tricarboxylic acid (TCA) cycle and the glyoxylate shunt in

Saccharomyces cerevisiae to succinic acid production during dough fermentation. *Int. J. Food Microbiol.* 204: 24–32.

35 Okino, S., Noburyu, R., Suda, M. et al. (2008). An efficient succinic acid production process in a metabolically engineered *Corynebacterium glutamicum* strain. *Appl. Microbiol. Biotechnol.* 81 (3): 459–464.

36 Shen, N.-k., Liao, S.-m., Wang, Q.-y. et al. (2016). Economical succinic acid production from sugarcane juice by *Actinobacillus succinogenes* supplemented with corn steep liquor and peanut meal as nitrogen sources. *Sugar Tech.* 18 (3): 292–298.

37 Chen, P.C., Zheng, P., Ye, X.Y., and Ji, F. (2017). Preparation of *A. succinogenes* immobilized microfiber membrane for repeated production of succinic acid. *Enzyme Microb. Technol.* 98: 34–42.

38 Yan, Q., Zheng, P., Tao, S.T., and Dong, J.J. (2014). Fermentation process for continuous production of succinic acid in a fibrous bed bioreactor. *Biochem. Eng. J.* 91: 92–98.

39 Jiang, M., Xu, R., Xi, Y.L. et al. (2013). Succinic acid production from cellobiose by *Actinobacillus succinogenes. Bioresour. Technol.* 135: 469–474.

40 Joshi, R.V., Schindler, B.D., McPherson, N.R. et al. (2014). Development of a markerless knockout method for *Actinobacillus succinogenes. Appl. Environ. Microbiol.* 80 (10): 3053–3061.

41 Shen, N., Zhang, H., Qin, Y. et al. (2018). Efficient production of succinic acid from duckweed (*Landoltia punctata*) hydrolysate by *Actinobacillus succinogenes* GXAS137. *Bioresour. Technol.* 250: 35–42.

42 Cao, W., Wang, Y., Luo, J. et al. (2018). Succinic acid biosynthesis from cane molasses under low pH by *Actinobacillus succinogenes* immobilized in luffa sponge matrices. *Bioresour. Technol.* 268: 45–51.

43 Ahn, J.H., Bang, J., Kim, W.J., and Lee, S.Y. (2017). Formic acid as a secondary substrate for succinic acid production by metabolically engineered *Mannheimia succiniciproducens. Biotechnol. Bioeng.* 114 (12): 2837–2847.

44 Nghiem, N.P., Montanti, J., and Kim, T.H. (2016). Pretreatment of dried distiller grains with solubles by soaking in aqueous ammonia and subsequent enzymatic/dilute acid hydrolysis to produce fermentable sugars. *Appl. Biochem. Biotechnol.* 179 (2): 237–250.

45 Cui, Z., Gao, C., Li, J. et al. (2017). Engineering of unconventional yeast Yarrowia lipolytica for efficient succinic acid production from glycerol at low pH. *Metab. Eng.* 42: 126–133.

46 Wahl, S.A., Bernal Martinez, C., Zhao, Z. et al. (2017). Intracellular product recycling in high succinic acid producing yeast at low pH. *Microb. Cell Fact.* 16 (1): 1–13.

47 Uysal, U. and Hamamcı, H. (2021). Succinic acid production from cheese whey via fermentation by using alginate immobilized *Actinobacillus succinogenes*. *Bioresour. Technol. Rep.* 16 (1): 100829.

48 Wang, C., Ming, W., Yan, D. et al. (2014). Novel membrane-based biotechnological alternative process for succinic acid production and chemical synthesis of bio-based poly (butylene succinate). *Bioresour. Technol.* 156: 6–13.

49 Babaei, M., Tsapekos, P., Alvarado-Morales, M. et al. (2019). Valorization of organic waste with simultaneous biogas upgrading for the production of succinic acid. *Biochem. Eng. J.* 147: 136–145.

50 Gadkari, S., Kumar, D., Qin, Z.h. et al. (2021). Life cycle analysis of fermentative production of succinic acid from bread waste. *Waste Manag.* 126: 861–871.

51 Isar, J., Agarwal, L., Saran, S., and Saxena, R.K. (2006). Succinic acid production from *Bacteroides fragilis*: process optimization and scale up in a bioreactor. *Anaerobe* 12 (5–6): 231–237.

52 Kuglarz, M., Alvarado-Morales, M., Dąbkowska, K., and Angelidaki, I. (2018). Integrated production of cellulosic bioethanol and succinic acid from rapeseed straw after dilute-acid pretreatment. *Bioresour. Technol.* 265: 191–199.

53 Chen, X., Wu, X., Jiang, S., and Li, X. (2017). Influence of pH and neutralizing agent on anaerobic succinic acid production by a Corynebacterium crenatum strain. *J. Biosci. Bioeng.* 124 (4): 439–444.

54 Jansen, M.L.A. and van Gulik, W.M. (2014). Towards large scale fermentative production of succinic acid. *Curr. Opin. Biotechnol.* 30: 190–197.

55 Hermann, T., Reinhardt, J., Yu, X., Udani, R., and Staples, L. (2012) Improved fermentation process for the production of organic acids. WO Patent 2012, 018699 A2, issued 9 February 2012.

56 This, A., Ridge, O., and Plus, C. (2002). Direct capture of products from biotransformations. *Proteins* 1–16.

57 Carnevali, D., Rigamonti, M.G., Tabanelli, T. et al. (2018). Levulinic acid upgrade to succinic acid with hydrogen peroxide. *Appl. Catal. A Gen.* 563: 98–104.

58 Dunlop, A.P. and Smith, S. (1954). Preparation of succinic acid, Patent: US2676186A, issued 20 April 1954. 1–2.

59 Wang, J., Zeng, A.-p., and Yuan, W. (2022). Succinic acid fermentation from agricultural wastes: the producing microorganisms and their engineering strategies. *Curr. Opin. Environ. Sci. Heal.* 25: 100313.

60 Zeikus, J.G., Jain, M.K., and Elankovan, P. (1999). Biotechnology of succinic acid production and markets for derived industrial products. *Appl. Microbiol. Biotechnol.* 51 (5): 545–552.

61 Saxena, R.K., Saran, S., Isar, J., and Kaushik, R. (2016). Production and applications of succinic acid. *Curr. Dev. Biotechnol. Bioeng. Prod. Isol. Purif. Ind. Prod.* 601–630.

62 Samuelov, N.S., Datta, R., Jain, M.K., and Zeikus, J.G. (1999). Whey fermentation by *Anaerobiospirillum succiniciproducens* for production of a succinate-based animal feed additive. *Appl. Environ. Microbiol.* 65 (5): 2260–2263.

63 Wang, Y., Xu, Y., Qin, X. et al. (2021). Effects of S,S-ethylenediamine disuccinic acid on the phytoextraction efficiency of *Solanum nigrum* L. and soil quality in Cd-contaminated alkaline wheat soil. *Environ. Sci. Pollut. Res.* 28 (31): 42959–42974.

64 Li, C., Ong, K.L., Cui, Z. et al. (2021). Promising advancement in fermentative succinic acid production by yeast hosts. *J. Hazard. Mater.* 401: 123414. (1–16).

65 Bozell, J.J., Moens, L., Elliott, D.C. et al. (2000). Production of levulinic acid and use as platform chemical for derived products. *Resour Conserv Recycl.* 28: 685–699.

66 Kawakami, H., Ebata, T., and Matsushita, H. (1991). A new synthesis of 5-aminolevulinic acid. *Agric. Biol. Chem.* 55 (6): 1687–1688.

67 Takeya, H., Shimizu, T., and Ueki, H. (1995). Process for preparing 5-aminolevulinic acid, Patent. US5380935A, issued 10 January 1995.

68 Bellardita, M., Virtù, D., Di Franco, F. et al. (2022). Heterogeneous photocatalytic aqueous succinic acid formation from maleic acid reduction. *Chem. Eng. J.* 431 (P2): 134131.

69 Nikolaevna, N., Ivanovich, S., Borisovich, J., and Mikhailovna, G. (1974). Method of Producing Zeta-aminolevulinic Acid Hydrochloride. United States Patent (19). (19), 2–4, issued 5 November 1974.

70 Ebata, T., Kawakami, H., Matsumoto, K., Koseki, K., and Matsushita, H. (1994). Method of preparing an acid additional salt of delta- aminolevulinic acid. United States Patent (19). (19), issued 8 February 1994.

71 Kang, Z., Ding, W., Gong, X. et al. (2017). Recent advances in production of 5-aminolevulinic acid using biological strategies. *World J. Microbiol. Biotechnol.* 33: 200.

72 Sasaki, K., Ikeda, S., Nishizawa, Y., and Hayashi, M. (1987). Production of 5-aminolevulinic acid by photosynthetic bacteria. *J. Ferment. Technol.* 65 (5): 511–515.

73 Kang, Z., Wang, Y., Gu, P. et al. (2011). Engineering *Escherichia coli* for efficient production of 5-aminolevulinic acid from glucose. *Metab. Eng.* 13 (5): 492–498.

74 Zhang, J., Weng, H., Zhou, Z. et al. (2019). Engineering of multiple modular pathways for high-yield production of 5-aminolevulinic acid in *Escherichia coli*. *Bioresour. Technol.* 274: 353–360.

75 Ramzi, A.B., Hyeon, J.E., Kim, S.W. et al. (2015). 5-Aminolevulinic acid production in engineered *Corynebacterium glutamicum* via C5 biosynthesis pathway. *Enzyme Microb. Technol.* 81: 1–7.

76 Fu, W., Lin, J., and Cen, P. (2008). Enhancement of 5-aminolevulinate production with recombinant *Escherichia coli* using batch and fed-batch culture system. *Bioresour. Technol.* 99 (11): 4864–4870.

77 Li, T., Guo, Y.Y., Qiao, G.Q., and Chen, G.Q. (2016). Microbial synthesis of 5-aminolevulinic acid and its coproduction with polyhydroxybutyrate. *ACS Synth. Biol.* 5 (11): 1264–1274.

78 Lou, J.W., Zhu, L., Wu, M.B. et al. (2014). High-level soluble expression of the hemA gene from *Rhodobacter capsulatus* and comparative study of its enzymatic properties. *J. Zhejiang Univ. Sci. B* 15 (5): 491–499.

79 Yu, X., Jin, H., Liu, W. et al. (2015). Engineering *Corynebacterium glutamicum* to produce 5-aminolevulinic acid from glucose. *Microb. Cell Fact.* 14 (1): 1–10.

80 Zhang, J., Kang, Z., Ding, W. et al. (2016). Integrated optimization of the in vivo heme biosynthesis pathway and the in vitro iron concentration for 5-aminolevulinate production. *Appl. Biochem. Biotechnol.* 178 (6): 1252–1262.

81 Noh, M.H., Lim, H.G., Park, S. et al. (2017). Precise flux redistribution to glyoxylate cycle for 5-aminolevulinic acid production in *Escherichia coli*. *Metab. Eng.* 43: 1–8.

82 Feng, L., Zhang, Y., Fu, J. et al. (2016). Metabolic engineering of *Corynebacterium glutamicum* for efficient production of 5-aminolevulinic acid. *Biotechnol. Bioeng.* 113 (6): 1284–1293.

83 Rao, P.R., Sasikala, C., and Ramana, V. (1994). Synthesis in microorganisms. *Methods* 10 (5): 451–459.

84 Nishikawa, S. and Murooka, Y. 5-ALA: production by fermentation and agricultural and biomedical applications. *Biotechnol. Genet. Eng. Rev.* 1: 149–170.

85 Kang, Z., Zhang, J., Zhou, J. et al. (2012). Recent advances in microbial production of δ-aminolevulinic acid and vitamin B12. *Biotechnol. Adv.* 30 (6): 1533–1542.

86 Hara, K.Y., Saito, M., Kato, H. et al. (2019). 5-Aminolevulinic acid fermentation using engineered *Saccharomyces cerevisiae*. *Microb. Cell Fact.* 18 (1): 1–8.

87 Benedikt, E., Köst, H., Institut, B., and München, D.U. (1986). Synthesis of 5-aminolevulinic acid. *Zeitschrift fuer Naturforschung Teil B Anorganische Chemie Organische Chemie* 12: 1593–1594.

88 Liu, X., Zhu, L., Song, Q. et al. (2018). Effects of 5-aminolevulinic acid on the photosynthesis, antioxidant system, and α-bisabolol content of *Matricaria recutita. Not. Bot. Horti Agrobot. Cluj-Napoca* 46 (2): 418–425.

89 Xu, W.P., Chen, X.F., Guo, H.J. et al. (2021). Conversion of levulinic acid to valuable chemicals: a review. *J. Chem. Technol. Biotechnol.* 96 (11): 3009–3024.

90 Novotny, A. and Stummer, W. (2003). 5-Aminolevulinic acid and the blood-brain barrier – A review. *Med. Laser Appl.* 18 (1): 36–40.

91 Elliott, J.T., Wirth, D.J., Davis, S.C. et al. (2021). Improving the usability of 5-aminolevulinic acid fluorescence-guided surgery by adding an optimized secondary light source. *World Neurosurg.* 149: 195–203.e4.

92 Hendawy, A.O., Khattab, M.S., Sugimura, S., and Sato, K. (2020). Effects of 5-aminolevulinic acid as a supplement on animal performance, iron status, and immune response in farm animals: a review. *Animals* 10 (8): 1–15.

93 Pileidis, F.D. and Titirici, M.M. (2016). Levulinic acid biorefineries: new challenges for efficient utilization of biomass. *ChemSusChem* 9 (6): 562–582.

94 Wang, T., Xu, H., He, J., and Zhang, Y. (2020). Investigation towards the reductive amination of levulinic acid by B (C 6 F 5) 3/hydrosilane system. *Tetrahedron* 76 (36): 131394.

95 Manzer, L.E. (2004). Production of 5-methyl-N-aryl-2- pyrroldone and 5-methyl-N-cycloalkyl-2-pyrrolidone by reductive amination of levulinc acid with arylamines. US 6743819 B1. US 6743819 B1, issued 2004.

96 Moreno-marrodan, C., Liguori, F., and Barbaro, P. (2019). Sustainable processes for the catalytic synthesis of safer chemical substitutes of N-methyl-2-pyrrolidone. *Mol. Catal.* 466: 60–69.

97 Bukhtiyarova, M.V. and Bukhtiyarova, G.A. (2021). Reductive amination of levulinic acid or its derivatives to pyrrolidones over heterogeneous catalysts in the batch and continuous flow reactors: a review. *Renew. Sustainable Energy Rev.* 143: 110876.

98 Manzer, L.E. (2004) Production of 5-methyl-N-aryl-2- pyrrolidone and 5-methyl-N-alkyl-2-pyrrolidone by reductive amination of levulinic acid with nitro. US 6818593 B2. US 6818593 B2, issued 2004.

99 Manzer, L.E. (2006) Process for converting cangelica lactone to 5-Methyl-N-alkyl-2-pyrrolidone using alkyl amines. US 7030249 B2. US 7030249 B2, issued 2006.

100 Gao, G., Sun, P., Li, Y. et al. (2017). Highly stable porous-carbon-coated Ni catalysts for the reductive amination of levulinic acid via an unconventional pathway. *ACS Catal.* 4927–4935.

101 Touchy, A.S., Siddiki, S.M.A.H., Kon, K., and Shimizu, K. (2014). Heterogeneous Pt catalysts for reductive amination of levulinic acid to pyrrolidones. *ACS Catal.* 4: 3045–3050.

102 Xie, C., Song, J., Wu, H. et al. (2019). Ambient reductive amination of levulinic acid to pyrrolidones over Pt nanocatalysts on porous TiO_2 nanosheets. *J. Am. Chem. Soc.* 141: 4002–4009. https://doi.org/10.1021/jacs.8b13024.

103 Zhang, J., Xie, B., Wang, L. et al. (2016). Zirconium oxide supported palladium nanoparticles as a highly efficient catalyst in the hydrogenation – amination of levulinic acid to pyrrolidones. *ChemCatChem* 9: 1–8.

104 Muzzio, M., Yu, C., Lin, H. et al. (2019). *Green Chem.* 21: 1895–1899. https://doi.org/10.1039/C9GC00396G.

105 Du, X., He, L., Zhao, S. et al. (2011). Hydrogen-independent reductive transformation of carbohydrate biomass into γ-valerolactone and pyrrolidone derivatives with supported gold catalysts. *Angew. Chem. Int. Ed.* 50: 7815–7819.

106 Martínez, J.J., Silva, L., Rojas, H.A. et al. (2017). Reductive amination of levulinic acid to different pyrrolidones on Ir/SiO_2-SO_3H: Elucidation of reaction mechanism. *Catal. Today* 296: 118–126. https://doi.org/10.1016/j.cattod.2017.08.038.

107 Wu, Y., Zhao, Y., Wang, H. et al. (2020). Ambient reductive synthesis of N-heterocyclic compounds over cellulose-derived carbon supported Pt nanocatalyst under H2 atmosphere. *Green Chem.* 22: 3820–3826. https://doi.org/10.1039/d0gc01177k.

108 Chieffi, G., Braun, M., and Esposito, D. (2015). Continuous reductive amination of biomass-derived molecules over carbonized filter paper-supported FeNi alloy. *ChemSusChem Commun.* 8: 3590–3594.

109 Wu, P., Li, H., and Fang, Z. (2022). Synergistic catalysis of Co-Zr/CN_x bimetallic nanoparticles enables reductive amination of biobased levulinic acid. *Adv. Sustainable Syst.* 6 (3): 2100321. (1 of 9).

110 Vidal, J.D., Climent, M.J., Concepción, P. et al. (2015). Chemoselective reductive amination of ethyl levulinate with amines Chemicals from biomass. *ACS Catal.*, https://doi.org/10.1021/acscatal.5b01113 5 (10): 5812–5821.

111 Wang, Y., Nuzhdin, A.L., Shamanaev, I.V. et al. (2022). Effect of phosphorus precursor, reduction temperature, and support on the catalytic properties of nickel phosphide catalysts in continuous-flow reductive amination of ethyl levulinate. *Int. J. Mol. Sci.* 23: 1106.

112 Wu, C., Zhang, H., Yu, B. et al. (2017). Lactate-based ionic liquid catalyzed reductive amination/cyclization of keto acids under mild conditions: a metal-free route to synthesize lactams. *ACS Catal.* 7: 7772–7776.

113 Bhujbal, A.V., Gokhale, T.A., and Bhanage, B.M. (2022). Reductive amination of biomass-based levulinic acid into pyrrolidone by protic ionic liquid via dehydrogenation of dimethyl amine borane. *Waste and Biomass Valorization* 13 (1): 443–451.

114 Siddiki, S.M.A.H., Touchy, A.S., Bhosale, A. et al. (2017). Direct synthesis of lactams from keto acids, nitriles, and H_2 by heterogeneous Pt catalysts. *ChemCatChem* 10 (4): 789–795.

115 Liu, Y., Zhang, K., Zhang, L. et al. (2021). Amination of levulinic acid/ester with nitriles over Pd/C. *React. Kinet. Mech. Catal.* 134 (2): 777–792.

116 Jouyban, A., Fakhree, M.A.A., and Shayanfar, A. (2010). Review of pharmaceutical applications of N-methyl-2-pyrrolidone. *J. Pharm. Pharm. Sci.* 13 (4): 524–535.

117 Fischer, K., Chen, J., Petri, M., and Gmehling, J. (2002). Solubility of H_2S and CO_2 in N-octyl-2-pyrrolidone and of H_2S in methanol and benzen. *AIChE J.* 48 (4): 887–893.

118 Shorvon, S. (2001). New drugs classes pyrrolidone derivatives. *Lancet* 358: 1885–1892.

119 Haaf, F., Sanner, A., and Straub, F. (1985). Polymers of N-vonylpyrrolidone: syntheses, characterization and uses. *Polym. J.* 17: 143–152.

120 Chen, M., Ike, M., and Fujita, M. (2001). Acute toxicity, mutagenicity, and estrogenicity of bisphenol-A and other bisphenols. *Environ Toxicol.* 17: 80–86.

121 Bozell, J.J., Moens, L., and Renewable, N. Production of levulinic acid and use as a platform chemical for derived products. *Resour. Conserv. Recycl.* 28: 685–699.

122 Ertl, J., Cerri, E., Rizzuto, M., and Caretti, D. (2014). Natural derivatives of diphenolic acid as substitutes for bisphenol-A. *AIP Conf. Proc.* 1599: 326–329.

123 Davis, M.E. (1993). New vistas in zeolite and molecular sieve catalysis. *Acc. Chem. Res.* 26 (3): 111–115.

124 Yu, X., Guo, Y., Li, K. et al. (2008). Catalytic synthesis of diphenolic acid from levulinic acid over cesium partly substituted Wells-Dawson type heteropolyacid. *J. Mol. Catal. A Chem.* 290 (1–2): 44–53.

125 Van de Vyver, S., Geboers, J., Helsen, S. et al. (2012). Thiol-promoted catalytic synthesis of diphenolic acid with sulfonated hyperbranched poly(arylene oxindole)s. *Chem. Commun.* 48 (29): 3497–3499.

126 Liu, Y., Zhang, Y., and Fang, Z. (2012). Design, synthesis, and application of novel flame retardants derived from biomass. *BioResources* 7 (4): 4914–4925.

127 Larrechi, M.S., Lligadas, G., Ronda, J.C. et al. (2011). Polybenzoxazines from renewable diphenolic acid. *J. Polym. Sci., Part A: Polym. Chem.* 49: 1219–1227.

128 Yue, L., Maiorana, A., Khelifa, F. et al. (2018). Surface-modi fi ed cellulose nanocrystals for biobased epoxy nanocomposites. *Polymer (Guildf)* 134: 155–162.

129 Varghai, D., Maiorana, A., Meng, Q. et al. (2016). Sustainable, electrically-conductive bioepoxy nanocomposites. *Polymer (Guildf)* 107: 292–301.

130 Maiorana, A., Spinella, S., and Gross, R.A. (2015). Bio-based alternative to the diglycidyl ether of bisphenol A with controlled materials properties. *Biomacromolecules* 16: 1021–1031.

131 Qian, Z., Xiao, Y., Zhang, X. et al. (2022). Bio-based epoxy resins derived from diphenolic acid via amidation showing enhanced performance and unexpected autocatalytic effect on curing. *Chem. Eng. J.* 435 (P2): 135022.

132 Bolton, E.K. (1942). Chemical Industry Medal. Development of Nylon. *Ind. Eng. Chem.* 34 (1): 53–58.

133 Qingxiang, G., Haifeng, L., Bing, L. et al. (2013). Diphenolic acid derivatives and preparation method and application thereof, CN1030588831A patent, issued 24 April 2013.

134 Qingxiang, G., Haifeng, L., Li, D. et al. (2012). Condensation of phenol with ethyl levulinate in acidic 1-n-alkyl-3-methylomodazolium ionic liquids. CN102701955B patent, issued 25 June 2014.

135 Weber, M. and Weiser-Elbl, K. (1998). Thermoplastic molding compositions and their use. European patent 855430 to BASF A.-G., issued 19 November 1987.

136 Heilig, M.L. (1994). United States patent office. *ACM SIGGRAPH Comput. Graph.* 28 (2): 131–134.

137 Chafetz, H., Liu, C.S., Papke, B.L, and Kenneedy, T.A. (1993). Lubricating oil composition containing the reaction product of an alkenylsuccinimide with a bis(hydroxyaromatic) substituted carboxylic acid. US005445750A patent, issued 29 August 1995.

138 Holmen, R.E. and Olander, S.J. (1977). Paint composition for marking paved surfaces. US patent 4031048 to Minnesota Mining and Mfg. Co., issued 21 June 1977.

139 Hatakeyama, J., Nagura, S., Motomi, K., Nagata, T., and Ishihara, T. (1988) Chemical amplifying type positive resist composition. US5750309A patent to Shin-Etsu Chemical Co, issued 12 May 1998.

140 Brooks, C.D.W., Bhatia, P., Kolasa, T. et al. (1997) Symmetrical bis-heteroarylmethoxyphenylalkyl carboxylates as inhibitors of leukotriene biosynthesis. CA2233550A1 patent to Abbott Laboratories, issued 10 April 1997.

141 Hoogeboom, T.J. (1974) Branched aromatic polycarbonate composition. US patent 3816373 to General Electric, issued 11 June 1974.

142 Malhotra, S.L., Naik, K.N., MacKinnon, D.N., and Jones, A.Y. (1997) Ink-jet printing sheet for transparency preparation. US patent 5683793 to Xerox, issued 4 November 1997.

143 Anderson, D., Frater, G., and Gygax, P. (1997) Fragrance precursors. World patent 9730687 to Givaudan-Roure (International) S.A., issued 28 August 1997.

144 Demmer, C.G. and Irving, E. (1986) Modified phenolic resins. European patent 184553 to Ciba-Geigy A.-G., issued 11 June 1986.

145 Rosenquist, N. (1991) Polycarbonate crosslinker resins and fire resistant compositions. European patent 372323 to General Electric Co, issued 29 January 1992.

6

Levulinic Acid in the Context of a Biorefinery

6.1 Biorefinery

A biorefinery may be conceived as an industrial plant that converts biomass or biomass-derived feedstock into biofuels, chemicals, and energy, keeping a high level of sustainability in the whole production process. In other words, a biorefinery is a CO_2 recycling unit, transforming sugars, triglycerides, and lignin that were produced in the photosynthesis process into valuable products and fuels (Figure 6.1). Ideally, a biorefinery would simultaneously produce biofuels, energy, and chemicals, but this arrangement is hardly ever seen in practice. Many biorefineries are focused on the production of biofuels, and some are designed to produce high added-value chemicals from biomass feedstock.

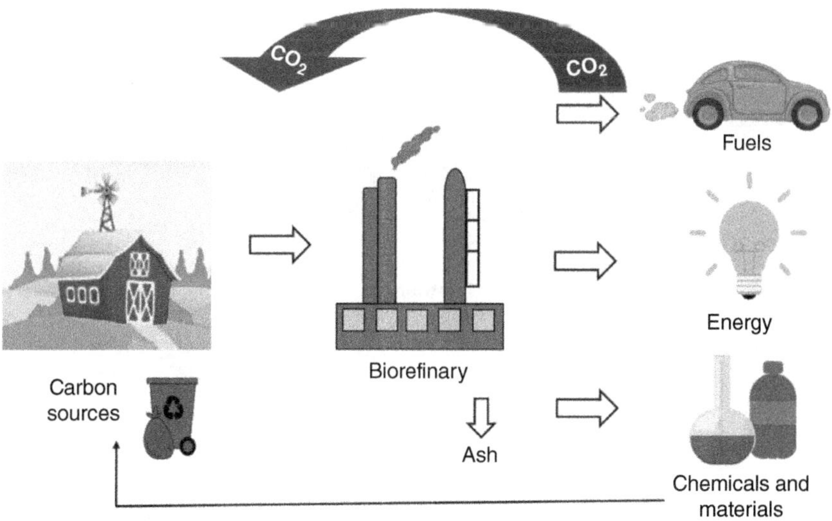

Figure 6.1 Schematic representation of a biorefinery.

Levulinic Acid: A Sustainable Platform Chemical for Value-Added Products, First Edition.
Claudio J.A. Mota, Ana Lúcia de Lima, Daniella R. Fernandes, and Bianca P. Pinto.

The most well-established biorefineries are based on sugar fermentation to produce bioethanol and on transesterification of triglycerides to yield biodiesel. A third type may be conceived for the production of cellulose and paper fibers, where lignocellulosic biomass is employed. Considering the production of levulinic acid (LA) and its derivatives, sugar-based biorefineries are the most promising, although lignocellulosic materials of the paper industry can also be valuable for this purpose. Nevertheless, the focus of this chapter will be mostly on sugar-based biorefineries trying to show examples of how the different products of this type of biorefinery can be integrated to produce levulinic acid and some of its derivatives.

6.2 Sugar-Based Biorefinery

Sugar-based biorefineries are mostly focused on the production of bioethanol through carbohydrate fermentation. In addition, sucrose and fructose syrup can be obtained when sugar cane and corn are used as feedstocks, respectively (Figure 6.2). In Brazil, sugar cane is the main feedstock for the production of bioethanol, whereas corn is predominant in the United States, although Brazil is also advancing in the corn biorefineries for the production of bioethanol too.

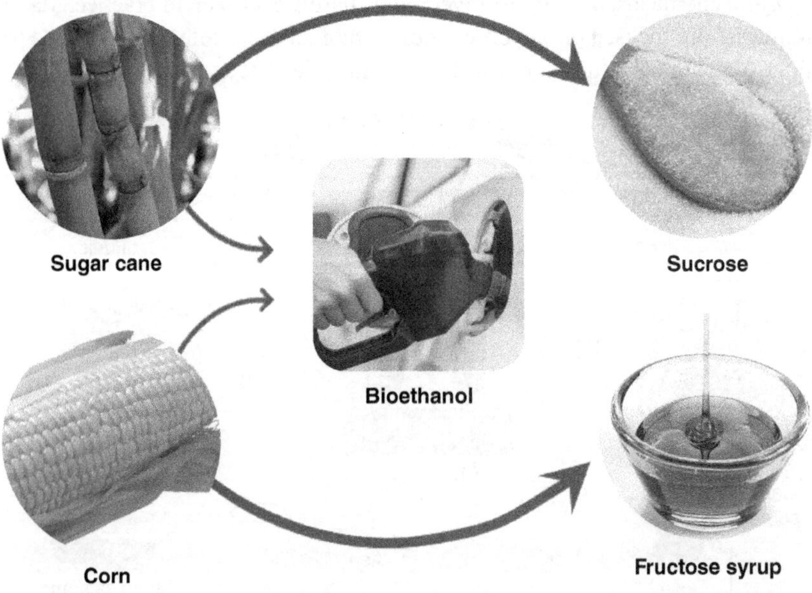

Figure 6.2 Main products obtained in sugar cane and corn-based biorefineries. Source: Ilton Rogerio/Adobe Stock, BillionPhotos.com/Adobe Stock, Maridav/Adobe Stock, vvoe/Adobe Stock, and Mara Zemgaliete/Adobe Stock.

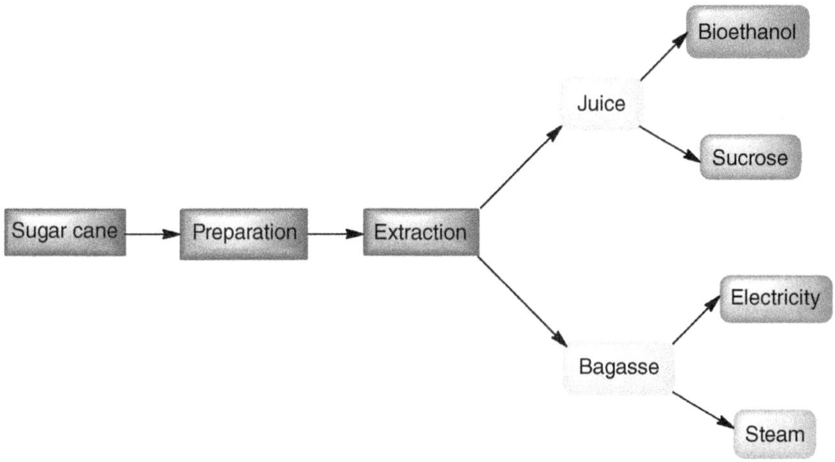

Figure 6.3 Simplified scheme of a sugar cane biorefinery to produce bioethanol, sucrose, and energy.

The sugar cane juice contains about 18% of sucrose. Figure 6.3 shows a simplified scheme of the production process. The juice is obtained together with the bagasse, which is normally burned to generate electricity and steam. The juice may be directed to produce sucrose, upon evaporation and recrystallization, as well as ethanol, upon fermentation and distillation. Therefore, the sugar cane biorefinery produces a biofuel (bioethanol), a chemical (sucrose), and energy (steam and electricity), being a good example of process integration.

The corn kernel composition consists of around 70% starch, which is a polysaccharide. In a very summarized way, the corn grains are milled and grinded to obtain the starch, which is then hydrolyzed with the aid of enzymes to yield a sugar-rich solution, mostly glucose, that could be fermented to ethanol. Alternatively, the sugar-rich solution may be treated with isomerase enzymes to yield what is known as high-fructose corn syrup (HFCS), which contains about 42% of fructose, and it is normally used in the food and beverage industry (Figure 6.4).

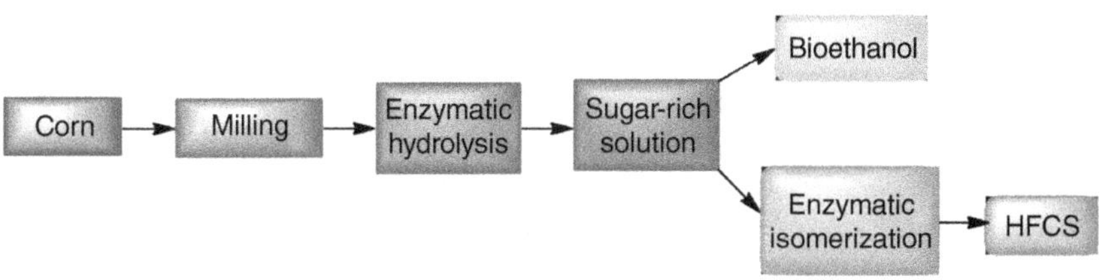

Figure 6.4 Simplified scheme of a corn-based biorefinery to produce bioethanol and high-fructose corn syrup (HFCS).

In recent years, the technology of second-generation ethanol, also known as cellulosic ethanol, has emerged as a viable alternative to increase the production of this biofuel [1]. The process involves the hydrolysis of cellulose and hemicellulose present in the biomass, affording glucose and other simple sugar molecules that can be converted to ethanol upon fermentation. In the case of a sugar cane biorefinery, the cellulosic ethanol can be obtained through the hydrolysis of the bagasse, obtained upon the juice extraction step. For the corn biorefinery, the cellulosic ethanol could come from the corn straw or the corn fiber, which are rich in lignocellulosic material.

6.3 Levulinc Acid and Levulinates from a Sugar Cane Biorefinery

As pointed out in Chapter 2, glucose and fructose can be converted into levulinic acid through acid treatment, with the intermediacy of 5-hydroxymethyl furfural (5-HMF). Therefore, the juice extracted from sugar cane is a valuable source of these carbohydrates. Levulinic acid could then be produced using some of the technologies described in Chapter 2 (Figure 6.5). In addition, ethyl levulinate could also be produced, either by traditional esterification with ethanol or from ethanolysis of the sugar molecules.

Sucrose is one of the cheapest bioderived chemicals, used worldwide as a natural sweetener in the food and beverage industry. The FOB price of sucrose in Brazil is, presently, around USD 0.50 per kilogram [2]. Thus, transforming it into levulinic acid would significantly add value, as the prices of levulinic acid range, at least, between 5- to 10-folds higher. Other derivatives that could be produced in a sugar cane biorefinery are the inorganic levulinates, particularly calcium and sodium levulinates. These chemicals find applications in the pharmaceutical, cosmetic, and food sectors, also showing high-added values.

Ethyl levulinate finds application as flavoring agent in the food and cosmetic sectors. The production costs would be significantly reduced if integration with a sugar cane biorefinery is considered, as both feedstocks, sucrose, and ethanol,

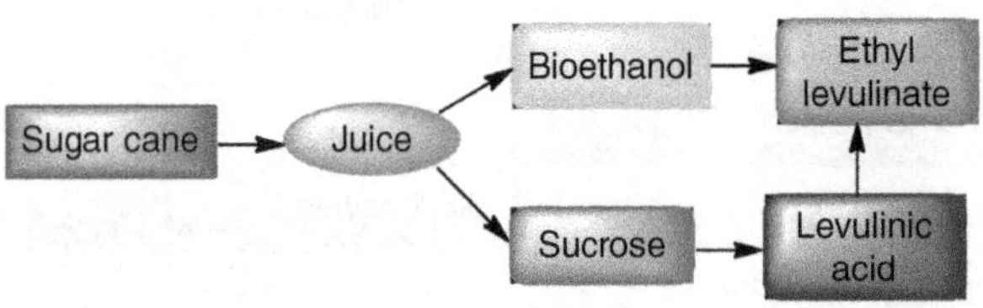

Figure 6.5 Simplified flow diagram of the production of levulinic acid and ethyl levulinate in a sugar cane biorefinery.

are the ultimate products. Thus, transforming sucrose into levulinic acid, through some acid hydrolysis technology, followed by acid-catalyzed esterification with ethanol would afford ethyl levulinate. Since ethyl levulinate is still considered a specialty chemical, its price may significantly vary depending on the producer, purity, and the amount purchased. Today, many sellers consider a price ranging from US$ 20 to 35 per kg. However, with developments of high-scale production plants, aiming to use ethyl levulinate in the fuel sector, the price is expected to decrease to cents of a dollar per kg of product [3], depending on the region, biomass feedstock, and technology employed. Hence, in the short-term, a sugar cane biorefinery would profit from selling ethyl levulinate as a specialty chemical, whereas in the long-term, it may provide this compound for the fuel sector, particularly for diesel-derived engines.

6.4 Production of γ-Valerolactone in a Sugar Cane Biorefinery

Hydrogenation of levulinic acid may afford γ-valerolactone (GVL), a chemical with application as solvent and fuel additive, as well as an intermediate for the production of higher hydrocarbons.

Most sugar cane biorefineries in Brazil burn the bagasse to produce electricity and steam for energy supply. Today, this electricity is used to run the plants, but a great part is directed to the grid and provides additional income to the plant, especially in times when the electricity price is high in the region. A third option that may be envisaged is the use of electricity to produce hydrogen through water electrolysis, a well-known technology, which in turn can be used to produce GVL (Figure 6.6). As the sugar cane biorefineries plants are usually near the plantation areas, there is no hydrogen supplier nearby, making its on-site production necessary.

Another possibility for the production of GVL from LA on a sugar cane biorefinery is the coutilization of the formic acid produced upon the transformation of fructose and glucose into LA. As mentioned in Chapter 4, there are many

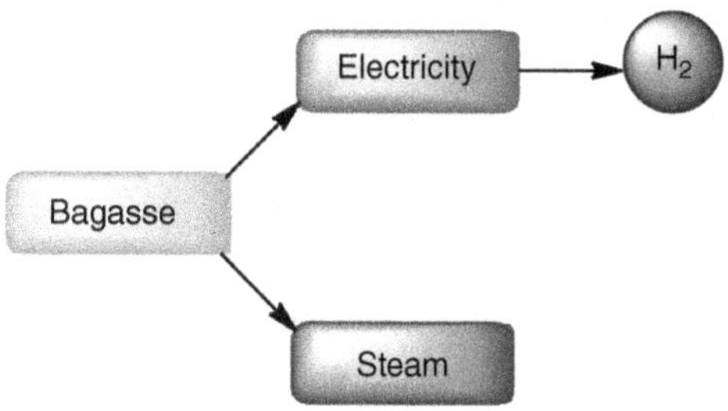

Figure 6.6 Schematic production of hydrogen on a sugar cane biorefinery.

$$HCO_2H \longrightarrow CO + H_2O \quad \Delta H = 7\ \text{kcal mol}^{-1} \quad \Delta S = 33.2\ \text{cal mol K}^{-1}$$

$$HCO_2H \longrightarrow CO_2 + H_2 \quad \Delta H = 7.5\ \text{kcal mol}^{-1} \quad \Delta S = 51.6\ \text{cal mol K}^{-1}$$

Figure 6.7 Formic acid decomposition pathways: dehydration and dehydrogenation.

metallic catalysts that decompose formic acid into hydrogen and CO_2, avoiding the direct use of hydrogen in the process. This approach would avoid the production of hydrogen in the plant, possibly reducing the capital and operational expenditures.

The decomposition of formic acid can proceed through two different pathways (Figure 6.7) [4]. Dehydration mostly occurs on the gas phase or in the presence of strongly acidic solutions. On the other hand, dehydrogenation is the preferred route in aqueous solutions and presence of metallic catalysts [4, 5]. Both reactions are endothermic, but present large positive entropy variation. Thus, the $\Delta G°$ values at 25 °C are – 3.0 kcal mol^{-1} and – 7.9 kcal mol^{-1} for dehydration and dehydrogenation, respectively. Therefore, both processes are favored at room temperature, which is quite unusual for endothermic reactions.

The selectivity of the catalytic system is of prime importance because CO is a poison for many metallic catalysts. Therefore, its concentration must be kept to a minimum to avoid rapid deactivation of the metal catalyst. Usually, the selectivity ratio between dehydrogenation and dehydration should be over 10^5 [6]. In general, heterogeneous metal catalyst systems are less selective than homogeneous systems and, therefore, yield high CO concentrations, above 1000 ppm, which may impact the long-term activity of the catalyst. The use of AgPd alloy nanoparticles supported on graphene oxide decompose formic acid to hydrogen without virtually producing CO as impurity [7, 8], but this type of catalyst is still far from being commercially used.

Despite the problem of CO production, LA hydrogenation with formic acid as a hydrogen source is widely studied [9]. Table 6.1 shows some literature results of LA hydrogenation with formic acid as hydrogen source in the liquid phase. The yield of GVL depends on the type of catalyst, formic acid/levulinc acid molar ratio, as well as on the reaction conditions.

Coke deposition and metal leaching are the major causes of catalyst deactivation. As the temperature increases, the rate of side reactions goes up decreasing the overall yield of GVL. Excessive FA/LA molar ratios may favor the hydrogenolysis of GVL to valeric acid.

The GVL market is still restricted. It is mainly used as flavoring and special solvent, including biomass pretreatment [14, 15]. Thus, in the short-term the

Table 6.1 LA hydrogenation to GVL using formic acid (FA) as hydrogen source and heterogeneous metal catalysts.

Catalyst	FA/LA[a)]	Solvent	T (°C)	t (min)	P (bar)	GVL Yield (%)	References
Ru/C	1	Water	150	5	AP[b)]	21	[10]
Ru/C	2	Water	150	5	AP[b)]	37	[10]
Ru/C	3	Water	150	5	AP[b)]	90	[10]
Ag/ZrO_2	1	Water	220	5	10	22	[11]
Ni/ZrO_2	1	Water	220	5	10	34	[11]
$Ag\text{-}Ni/ZrO_2$	1	Water	220	5	10	99	[11]
Ni/Al_2O_3	5	Water	250	1	10	89	[12]
Ni/HT[c)]	5	Water	250	1	10	48	[12]
Ru/ZrO_2	1	Water	150	12	10	73	[13]
Ru/TiO_2	1	Water	150	12	10	16	[13]

a) Formic acid/levulinic acid molar ratio.
b) AP accounts for autogeneous pressure.
c) HT accounts for hydrotalcite.

production of GVL in a sugar cane biorefinery may be for internal use, to help the process of second-generation ethanol from hexoses produced from the hydrolysis of cellulose and hemicellulose of the bagasse, because GVL has been shown to be a good solvent for biomass pretreatment.

The use of GVL to assist in the conversion of lignocellulosic materials may be applied to different biomasses, not being restricted to sugar cane bagasse. GVL has excellent solvent properties, with melting point of−31 °C and boiling point of 207 °C, besides being nontoxic. In general, GVL is used in mixtures with water and mineral acid, at temperatures above 100 °C, to deconstruct lignocellulosic biomass materials. For instance, treatment of hardwood with 80% aqueous GVL solution containing diluted H_2SO_4 led to 80% of lignin removal and yielded 99% and 96% of the original glucan and xylan, respectively [15]. Liquid CO_2 at high pressure was used to recover the GVL (Figure 6.8).

GVL mixed with water, DMSO, or DMF can be used to efficiently solubilize lignin [16]. The lignin can be recovered upon addition of ethanol. Aqueous GVL acidified with H_2SO_4 or HCl can directly convert corn stover, hardwood, and softwood into soluble carbohydrates, without the need for further enzymatic processing [17]. The carbohydrates can be recovered from the GVL fraction upon extraction with liquid CO_2 or addition of aqueous NaCl solutions. The yields are within 70–90%.

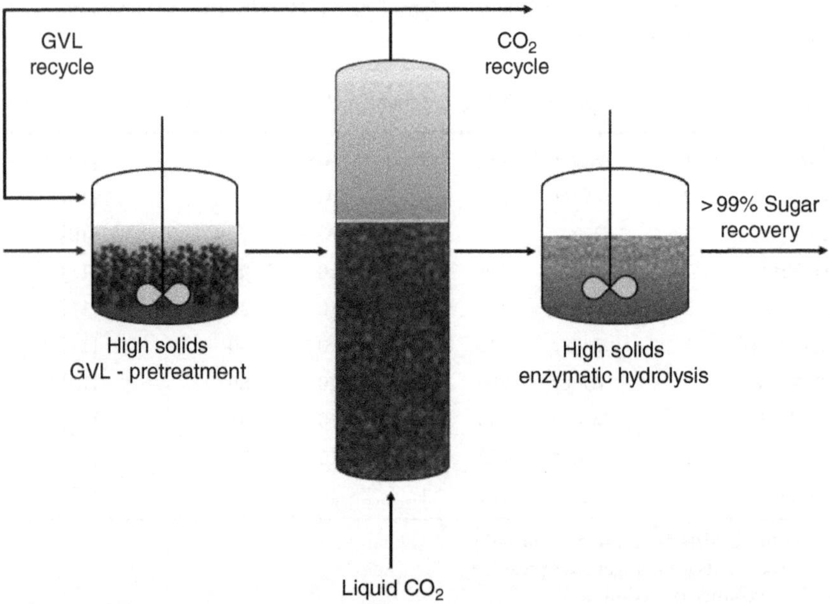

Figure 6.8 Schematic representation of hardwood pretreatment with GVL/water acidic solution, showing recovery with liquid CO_2.

6.5 LA in the Context of a Biodiesel Plant

Biodiesel is usually produced from the transesterification of vegetable oils and fats [18]. The triglycerides are converted into methyl, and in some cases, ethyl esters of the fatty acids (FAME) upon reaction with methanol or ethanol in the presence of a basic catalyst, also producing glycerol (Scheme 6.1).

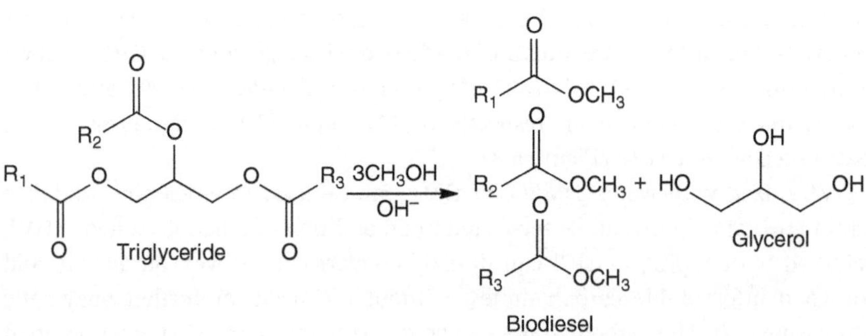

Scheme 6.1 Transesterification of triglycerides with methanol to produce biodiesel (FAME) and glycerol.

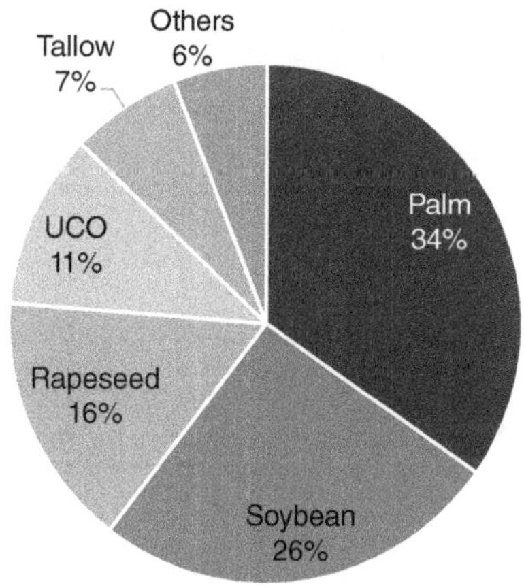

Figure 6.9 World share of feedstocks used for biodiesel production; reference year. Source: Adapted from UFOP Chart of the week (50 2019), press release, 2019-12 [19].

Today, soybean is the major crop for biodiesel production in Brazil, Argentina, and the United States, whereas palm is the main source of biodiesel in Asia, especially in Malaysia and Indonesia. In Europe, most biodiesel is produced from rapeseed. Figure 6.9 shows the global share of the vegetable oils used for the production of biodiesel [19].

Glycerol is obtained in approximately 10 wt% from the transesterification of triglycerides, but it is still not widely used worldwide. The chemical conversion of glycerol is promising [20] and some commercial processes have been implemented, highlighting the production of epichloridrin, a commodity used in the production of epoxy resins among other uses, and solketal, which is presently used as solvent but has potential to be used as fuel additive [21].

The integration of the biodiesel and levulinic acid chain productions could be made aiming at producing glycerol levulinic acid or ester ketals (GLEK), the process routes and uses of which have been highlighted in Chapter 5. This would add value to the glycerol of biodiesel production. The LA could be produced from the agricultural residues obtained upon the extraction of vegetable oils. For instance, the production of soybean oil usually involves pressing and solvent extraction, yielding what is called the soybean cake or soybean flour as residue. Today, most of the soybean flour is destinated for animal food supplements, as it is protein-rich. The average composition of the soybean flour is shown in Table 6.2. One can see that it presents significant amounts of carbohydrates, which can be converted into LA. Considering that the soy seed contains approximately 20 wt% of oil, the biomass cake generated upon oil extraction is a major residue that can be directed, at least in part, for the production of LA.

The simplified flow sheet of the integrated soybean biodiesel biorefinery with LA production is shown in Figure 6.10. Upon oil extraction, part of the soy cake can be used as feedstock for the production of levulinic acid and levulinate esters

Table 6.2 Average composition of soy flour (soy meal) per 100 g.

Component	g
Water	4.6
Proteins	49.8
Lipids (total fat)	8.9
Carbohydrates	30.6
Ashes	6.0

Source: USDA [22].

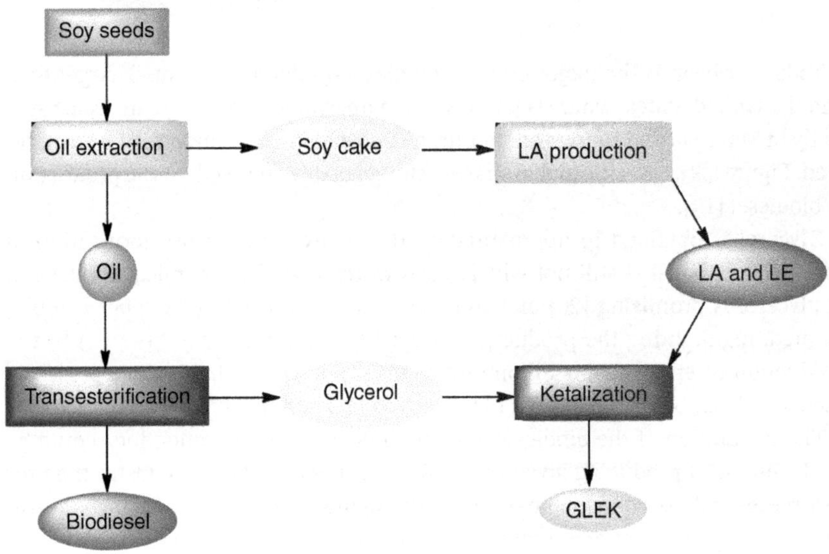

Figure 6.10 Simplified flow sheet for integrating the biodiesel and the LA chains in a soybean biorefinery.

(LE). The extracted oil is transformed into biodiesel upon transesterification, yielding glycerol as by-product. The integration of both production chains involves the reaction of LA or LE with glycerol, using acid catalysts, to afford GLEK.

6.6 Conclusions

Levulinic acid could be integrated into sugar cane and soybean biorefineries. While the sugar cane biorefinery can use sugar as well as lignocellulosic material to produce LA, the soy flour, obtained upon the extraction of soybean oil, may be the main feedstock for LA production in the latter.

The LA produced in sugar cane biorefineries could be transformed into ethyl levulinate, through reaction with the ethanol obtained from sugar fermentation. Another important derivative that could be produced is GVL, using hydrogen generated from water electrolysis with the energy generated from the burning of the bagasse.

The soybean biorefinery may integrate the biodiesel and LA production chains. Whereas the soybean oil may be used to produce biodiesel by transesterification, the soy cake residue obtained in the extraction of the vegetable oil may serve as feedstock to produce LA. The integration of both production chains can be accomplished using the glycerol formed in the transesterification of vegetable oil and the LA, or the correspondent methyl esters, to produce ketals of great commercial importance.

References

1 Aditiya, H.B., Mahlia, T.M.I., Chong, W.T. et al. (2016). Second generation bioethanol production: a critical review. *Renewable Sustainable Energy Rev.* 66: 631–653.

2 Sugar brazil fob Ports Indices. https://www.commodity3.com/chain/SUGBRINX/sugar-brazil-fob-ports-indices.

3 Leal Silva, J.F., Grekin, R., Mariano, A.P., and Maciel Filho, R. (2018). Making levulinic acid and ethyl levulinate economically viable: a worldwide technoeconomic and environmental assessment of possible routes. *Energy Technol.* 6 (4): 613–639.

4 Navlani-García, M., Mori, K., Salinas-Torres, D. et al. (2019). New approaches toward the hydrogen production from formic acid dehydrogenation over pd-based heterogeneous catalysts. *Front. Mater.* 6 (March): 1–18.

5 Tedsree, K., Li, T., Jones, S. et al. (2011). Hydrogen production from formic acid decomposition at room temperature using a Ag-Pd core-shell nanocatalyst. *Nat. Nanotechnol.* 6 (5): 302–307.

6 Eppinger, J. and Huang, K.W. (2017). Formic acid as a hydrogen energy carrier. *ACS Energy Lett.* 2 (1): 188–195.

7 Zhu, Q.L., Tsumori, N., and Xu, Q. (2015). Immobilizing extremely catalytically active palladium nanoparticles to carbon nanospheres: a weakly-capping growth approach. *J. Am. Chem. Soc.* 137 (36): 11743–11748.

8 Chen, Y., Zhu, Q.L., Tsumori, N., and Xu, Q. (2015). Immobilizing highly catalytically active noble metal nanoparticles on reduced graphene oxide: a non-noble metal sacrificial approach. *J. Am. Chem. Soc.* 137 (1): 106–109.

9 Yu, Z., Lu, X., Xiong, J. et al. (2020). Heterogeneous catalytic hydrogenation of levulinic acid to γ-valerolactone with formic acid as internal hydrogen source. *ChemSusChem* 13 (11): 2916–2930.

10 Feng, J., Gu, X., Xue, Y. et al. (2018). Production of γ-valerolactone from levulinic acid over a Ru/C catalyst using formic acid as the sole hydrogen source. *Sci. Total Environ.* 633: 426–432.

11 Hengne, A.M., Malawadkar, A.V., Biradar, N.S., and Rode, C.V. (2014). Surface synergism of an Ag-Ni/ZrO_2 nanocomposite for the catalytic transfer hydrogenation of bio-derived platform molecules. *RSC Adv.* 4 (19): 9730–9736.

12 Varkolu, M., Velpula, V., Burri, D.R., and Kamaraju, S.R.R. (2016). Gas phase hydrogenation of levulinic acid to γ-valerolactone over supported Ni catalysts with formic acid as hydrogen source. *New J. Chem.* 40 (4): 3261–3267.

13 Wang, Y., Liu, F., Han, H. et al. (2018). Metal phosphide: a highly efficient catalyst for the selective hydrodeoxygenation of furfural to 2-methylfuran. *ChemistrySelect* 3 (27): 7926–7933.

14 Raj, T., Chandrasekhar, K., Banu, R. et al. (2021). Synthesis of γ-valerolactone (GVL) and their applications for lignocellulosic deconstruction for sustainable green biorefineries. *Fuel* 303 (June): 121333.

15 Shuai, L., Questell-Santiago, Y.M., and Luterbacher, J.S. (2016). A mild biomass pretreatment using γ-valerolactone for concentrated sugar production. *Green Chem.* 18 (4): 937–943.

16 Xue, Z., Zhao, X., Sun, R.C., and Mu, T. (2016). Biomass-derived γ-valerolactone-based solvent systems for highly efficient dissolution of various lignins: dissolution behavior and mechanism study. *ACS Sustainable Chem. Eng.* 4 (7): 3864–3870.

17 Luterbacher, J.S., Rand, J.M., Alonso, D.M. et al. (2014). Nonenzymatic sugar production from biomass using biomass-derived γ-valerolactone. *Science* 343 (6168): 277–280.

18 Rezende, M.J.C., de Lima, A.L., Silva, B.V. et al. (2021). Biodiesel: an overview II. *J. Braz. Chem. Soc.* 32 (7): 1301–1344.

19 (2019). Global biodiesel production is increasing. https://renewable-carbon.eu/news/global-biodiesel-production-is-increasing/.

20 Mota, C.J.A., Pinto, B.P., and de Lima, A.L. (2017) *Glycerol: a versatile renewable feedstock for the chemical industry.*

21 Ozorio, L.P., Pianzolli, R., Mota, M.B.S., and Mota, C.J.A. (2012). Reactivity of glycerol/acetone ketal (solketal) and glycerol/formaldehyde acetals toward acid-catalyzed hydrolysis. *J. Braz. Chem. Soc.* 23 (5): 931–937.

22 (2019). Soy flour, low-fat. https://www.medindia.net/nutrition-data/soy-flour-low-fat.htm.

Index

Levulinic Acid: A Sustainable Platform Chemical for Value-Added Products, First Edition.
Claudio J.A. Mota, Ana Lúcia de Lima, Daniella R. Fernandes, and Bianca P. Pinto.